CONTEMPORARY'S
REAL NUMBERS
Developing Thinking Skills in Math
Geometry Basics

Allan D. Suter

Project Editor
Kathy Osmus

CONTEMPORARY BOOKS

CHICAGO

No part of this publication may be reproduced, stored in
a retrieval system, or transmitted in any form or by any
means, without the prior written permission of the
publisher.

Published by Contemporary Books, Inc.
180 North Michigan Avenue, Chicago, Illinois 60601
Manufactured in the United States of America
International Standard Book Number: 0-8092-4210-9

Published simultaneously in Canada by
Fitzhenry & Whiteside
195 Allstate Parkway
Valleywood Business Park
Markham, Ontario L3R 4T8
Canada

Editorial Director	*Production Editor*
Caren Van Slyke	Marina Micari
Editorial	*Cover Design*
Chris Benton	Lois Stein
Janet Fenn	
Ellen Frechette	*Illustrator*
Eunice Hoshizaki	Cliff Hayes
Laura Larson	
Steve Miller	*Art & Production*
Robin O'Connor	Carolyn Hopp
Seija Suter	
	Typography
Editorial Assistant	Impressions, Inc.
Erica Pochis	Madison, Wisconsin
Editorial Production Manager	
Norma Fioretti	

Cover photo © by Michael Slaughter and courtesy of
Century Tile, Itasca, Illinois.

CONTENTS

Geometric Shapes

Geometric shapes can be found in everyday objects. Everything from cars to radios to baseballs contains one or more geometric shapes.

▶ Find the shapes that are listed below in the drawings. Put an ✕ through an example of each shape for each drawing.

circle ◯

rectangle ▭

line segment ——

right angle ∟

point •

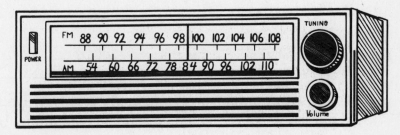

Geometric Ideas

Geometry studies the size, shape, and location of objects in space. Three basic concepts in geometry are **point, line,** and **line segment.**

A **point** is a location in space and can be shown by making a dot. Capital letters are used to name a point.

A **line** is a continuous set of points that extends forever in both directions. A line can be named by labeling any two points on the line.

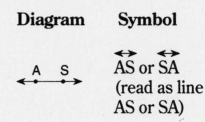

Diagram   **Symbol**

$\overleftrightarrow{AS}$ or $\overleftrightarrow{SA}$
(read as line AS or SA)

A **line segment** is part of a line with two endpoints. A line segment is named using both endpoints.

Diagram   **Symbol**

$\overline{DC}$ or $\overline{CD}$
(read as line segment DC or CD)

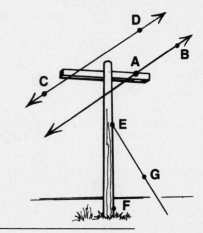

▶ Use the picture to list all points, lines, and line segments. Use symbols when needed.

1. Points: _G_ , _B_ , ___, ___, ___, ___, ___

2. Line segments: _GE_ , ___, ___, ___

3. Lines: $\overleftrightarrow{AB}$, ___

___

▶ We see geometry being applied every day. Describe as many points, lines, and line segments as you can using the labeled points below.

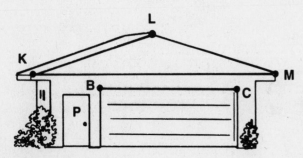

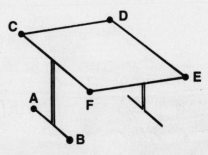

4. Points: ___, ___, ___, ___, ___, ___

5. Line segments: ___, ___, ___, ___

6. Points: ___, ___, ___, ___, ___, ___

7. Line segments: ___, ___, ___, ___, ___

Angles

An **angle** is formed when two lines meet at a point. The point where the lines meet is called the **vertex** of the angle. The lines that form the angle are called the **sides** of the angle. The symbol for an angle is ∠.

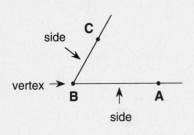

Example

∠ CBA (read as angle CBA) or ∠ ABC

- The first letter is always a point on one of the sides.
- The middle letter is always the vertex.
- The last letter is always a point on the other side.

▶ Name each angle two ways.

1.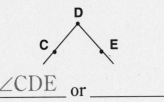

$\underline{\angle\text{CDE}}$ or _____

2.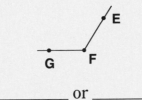

_____ or _____

3.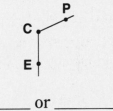

_____ or _____

4. Use the diagram to name six different angles.

a) $\underline{\angle\text{ANP}}$ d) _____

b) _____   e) _____

c) _____   f) _____

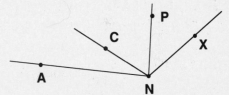

5. From the drawing, list eight different angles using the labeled points.

a) _____   e) _____

b) _____   f) _____

c) _____   g) _____

d) _____   h) _____

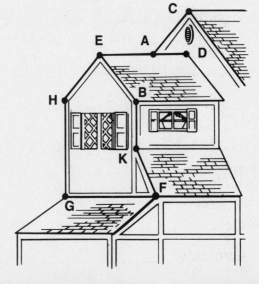

Geometric Models

▶ We see geometry and its applications everywhere. Describe as many points, lines, line segments, and angles as you can using the labeled points in the drawings below. Be sure to use symbols to label lines, angles, and line segments.

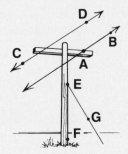

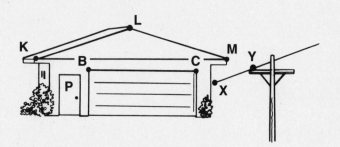

1. a) Points: _____

b) Line segments: $\overline{AB}$ _____

c) Angles: _____

d) Lines: _____

4. a) Points: _____

b) Line segments: _____

c) Angles: _____

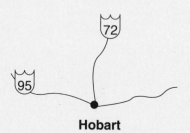

Hobart

2. Is Hobart a point, line segment, angle, or line? _____

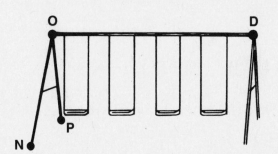

5. a) Points: _____

b) Line segments: _____

c) Angles: _____

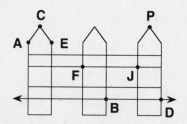

3. a) Points: _____

b) Line segments: _____

c) Angles: _____

6. a) Points: _____

b) Line segments: _____

c) Angles: _____

d) Lines: _____

Size of Angles

The size of an angle depends only on the "opening" between its sides. A large angle has a larger opening than a smaller angle.

Example A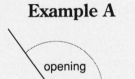

Example B

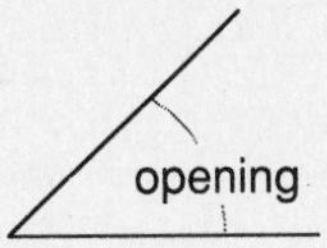

Example A shows a larger angle than Example B because the "opening" is larger.

▶ Circle the larger angle in each pair.

1. a) b)    **2.** a)  b)    **3.** a)  b)

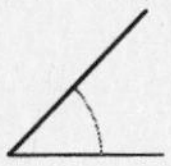

▶ Circle the smaller angle in each pair.

4. a) b)    **5.** a)  b)   **6.** a) b)

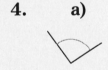

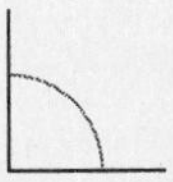

▶ Circle the two angles in each group that look about the same size.

7. a) b)  c)    **8.** a) 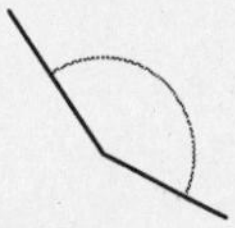 b) c)

▶ Draw a line to complete an angle that is about the same size as the given angle.

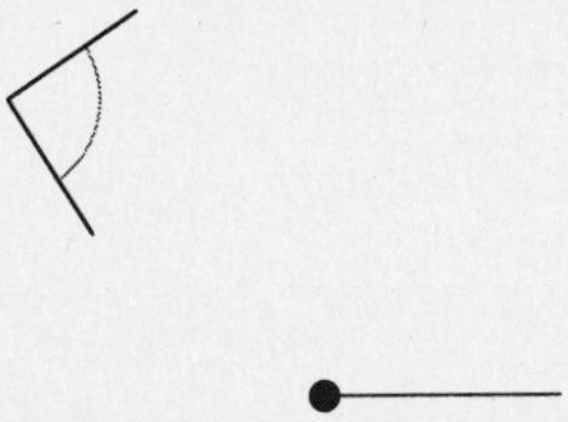

9.   **10.**   **11.**

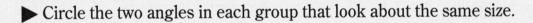

Using a Protractor

A **protractor** is a tool that measures the "opening" between two sides of an angle. The standard unit for measuring angles is the degree (°).

A protractor has two scales: an inside scale and an outside scale. Both are marked in degrees, beginning at 0° and ending at 180°.

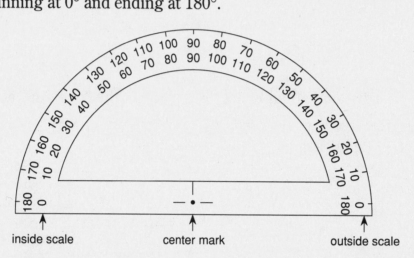

inside scale   center mark   outside scale

On this protractor, the inside scale starts at the left.   On this protractor, the outside scale starts at the right.

Example: Measure ∠ ABC.

Step 1: Line up the center mark on the protractor with the vertex of the angle.

Step 2: Line up the 0° mark with one side of the angle.

> **Note:** In this example, the side of the angle passes through the outside 0° mark. Read the outside scale.

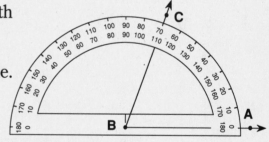

Step 3: Read the angle size. The size of ∠ ABC is 70°.

▶ Measure each angle.

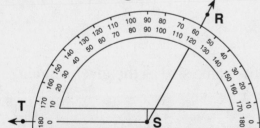

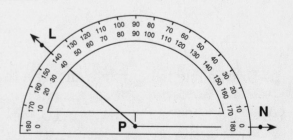

1. a) Will you use the inside or outside scale? _____

    b) ∠ RST measures _____

2. a) Will you use the inside or outside scale? _____

    b) ∠ LPN measures _____

Measuring Angles

▶ Use a protractor to measure each angle. You may need to extend the sides of the angle to measure the angle.

1. _____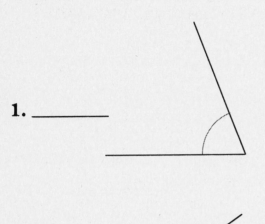

2. _____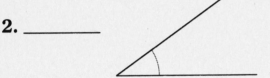

3. _____

4. _____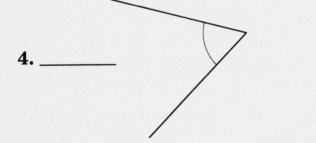

5. _____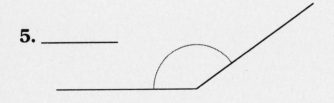

6. _____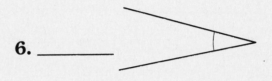

7. _____

8. _____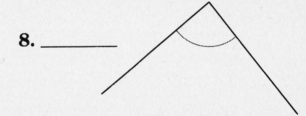

9. _____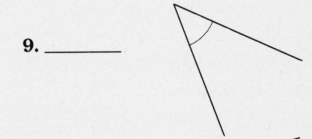

10. _____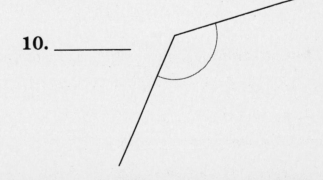

Classifying Angles

Angles are classified by size. Shown below are three ways to classify angles.

| The size of a **right angle** measures exactly 90°. | The size of an **acute angle** measures less than 90°. | The size of an **obtuse angle** measures more than 90°. |

90° This symbol shows a right angle. 0°   90° 0°   90° 180° 0°

▶ A corner of this paper measures 90° (a right angle). Use a corner of a sheet of paper to tell whether the following angles are > (greater than), < (less than), or = (equal to) a right angle. Then classify each angle as a right angle, an acute angle, or an obtuse angle.

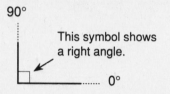 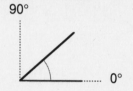 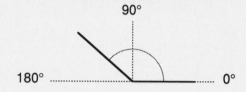

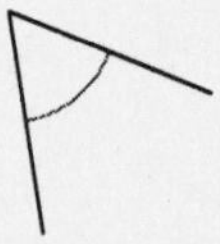

1. a) _____
    >, <, or =

b) _____ right
    classify

2. a) _____
    >, <, or =

b) _____
    classify

3. a) _____
    >, <, or =

b) _____
    classify

4. a) _____
    >, <, or =

b) _____
    classify

▶ Estimate the size of each angle without measuring. Then classify each angle.

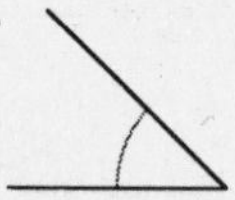

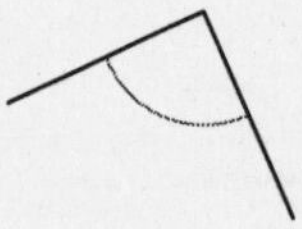

5. a) _____
    estimate

b) _____
    classify

6. a) _____
    estimate

b) _____
    classify

7. a) _____
    estimate

b) _____
    classify

8. a) _____
    estimate

b) _____
    classify

▶ Classify each angle as right, acute, or obtuse.

9. 32° _____   **11.** 90° _____   **13.** 170° _____

10. 118° _____   **12.** 12° _____   **14.** 89° _____

Complementary Angles

Two angles are **complementary** if the sum of their sizes is 90°.

Example A

Some complementary angles share a common side.

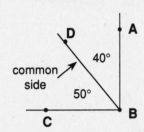

∠ ABD + ∠ DBC = 90°
The two angles are complementary.

Example B

Some complementary angles do not share a common side.

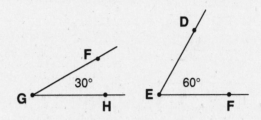

∠ FGH + ∠ DEF = 90°
The two angles are complementary.

▶ Find the size of the complementary angle for each angle.

1. _____

2. _____

3. _____

4. ∠ ABC = 85°
  ∠ CBD = _____

5. ∠ GOP = 46°
  ∠ DFA = _____

6. ∠ JMK = 8°
  ∠ NLJ = _____

7. ∠ RST = 29°
  ∠ GHI = _____

8. ∠ GWX = 60°
  ∠ LEV = _____

9. ∠ BMK = 53°
  ∠ CMJ = _____

▶ Draw the complement of each angle. Use the corner of a sheet of paper if you need help. The first one has been done for you.

10. a)   b)   c)   d)

Supplementary Angles

Two angles are **supplementary** if the sum of their sizes is 180°.

Example A

Some supplementary angles share a common side.

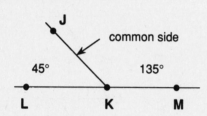

∠ JKL + ∠ JKM = 180°
The two angles are supplementary.

Example B

Some supplementary angles do not share a common side.

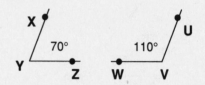

∠ XYZ + ∠ UVW = 180°
The two angles are supplementary.

▶ Find the size of the supplementary angle for each angle.

1. _____ 50° ?

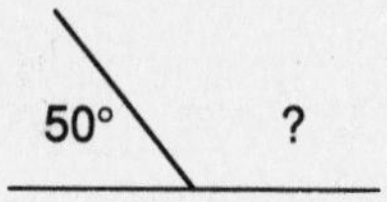

2. _____ 140° ?

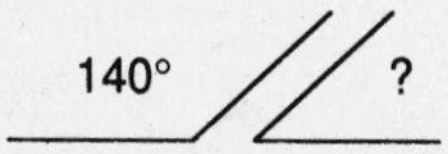

3. _____ ? 90°

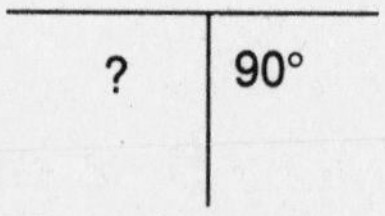

4. ∠ BHN = 20°

  ∠ AHP = _____

5. ∠ CDE = 154°

  ∠ FGJ = _____

6. ∠ RTV = 85°

  ∠ JKL = _____

7. ∠ FGW = 169°

  ∠ XGY = _____

8. ∠ MBC = 45°

  ∠ MBH = _____

9. ∠ CNO = 105°

  ∠ CBG = _____

▶ Draw the supplement of each angle. Use the straight edge of a ruler or a sheet of paper if you need help. The first one has been done for you.

10. a) 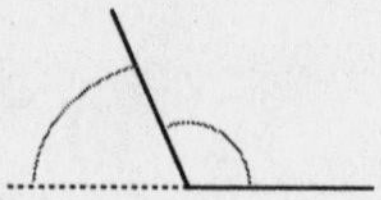   **b)**    **c)** 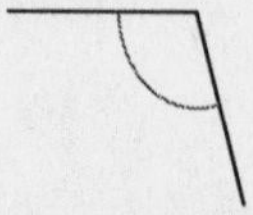   **d)**

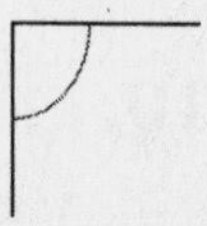

Vertical, Horizontal, and Slanting Lines

A line extends forever in opposite directions. Words such as *vertical, horizontal,* and *slanting* are used to describe the direction of a line.

A horizontal line runs from left to right. The line is level like the horizon.

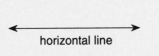

horizontal line

A vertical line runs up and down from the horizon.

vertical line

A slanting line is neither horizontal nor vertical.

slanting lines

▶ Answer the following questions.

1. Connect two points to form a horizontal line.

2. Connect two points to form a slanting line.

3. Connect two points to form a vertical line.

▶ Name (using symbols and letters) the horizontal, vertical, and slanting lines represented by letters below.

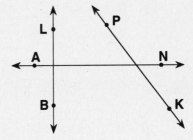

4. a) Horizontal: _____

   b) Vertical: _____

   c) Slanting: _____

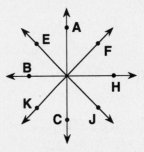

5. a) Horizontal: _____

   b) Vertical: _____

   c) Slanting: _____

Parallel and Intersecting Lines

▶ Two lines that have a common point are **intersecting lines**. Write the point that falls on both lines. Extend the lines if necessary.

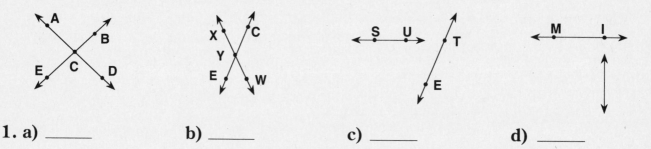

1. a) _____   **b)** _____   **c)** _____   **d)** _____

▶ Lines that do not intersect are **parallel lines.** Write *parallel* or *intersecting* for each pair of lines. Extend the lines in each problem to see if they will intersect.

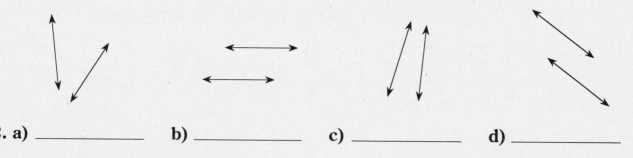

2. a) _____   **b)** _____   **c)** _____   **d)** _____

3. Draw two lines in each box that:

a) do not intersect   **b)** intersect at one point

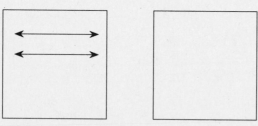

4. Draw a horizontal line and a vertical line in the box below. Are the lines intersecting or parallel? _____

5. Draw three lines in each box that:

a) do not intersect   **b)** intersect at one point   **c)** intersect at two points   **d)** intersect at three points

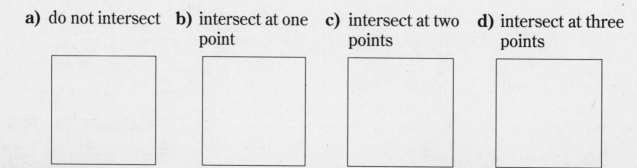

Perpendicular Lines

Two lines that intersect and form four right angles are called **perpendicular lines.**

Intersecting lines that **are** perpendicular:

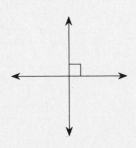

Intersecting lines that **are not** perpendicular:

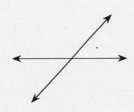

▶ Write *intersecting* or *perpendicular* for each pair of lines. Extend the lines if necessary. Use the corner of a sheet of paper if you need help checking for right angles.

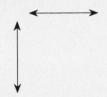   

1. a) _____ **b)** _____ **c)** _____ **d)** _____

▶ Line segments that intersect to form a right angle are perpendicular. Use the dots below to draw perpendicular line segments.

2.

3.

4.

5. Name eight pairs of perpendicular line segments from the picture below.

a) __$\overline{GF}$__ and __$\overline{BD}$__   **e)** _____ and _____

b) _____ and _____   **f)** _____ and _____

c) _____ and _____   **g)** _____ and _____

d) _____ and _____   **h)** _____ and _____

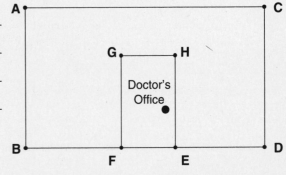

Street Plans

City maps of streets show:
- streets that intersect
- streets that are parallel
- streets that are perpendicular
- streets that form angles

▶ Use the street maps below to answer the questions.

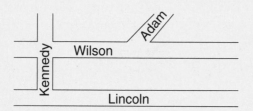

1. Which streets intersect?

   a) _____ and _____

   b) _____ and _____

   c) _____ and _____

2. Name the streets that are parallel.

   _____ and _____

3. Which streets are perpendicular?

   a) _____ and _____

   b) _____ and _____

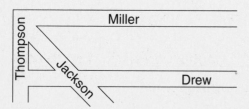

4. Which streets intersect with Miller?

   _____ and _____

5. If Miller Street and Drew Street continue as shown, will they intersect?

   _____

6. Which streets are perpendicular?

   a) _____ and _____

   b) _____ and _____

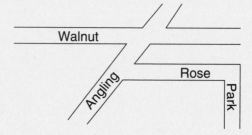

7. Which street is parallel to Rose?

   _____

8. Which street is perpendicular to Rose?

   _____

9. Which streets intersect with Rose?

   _____

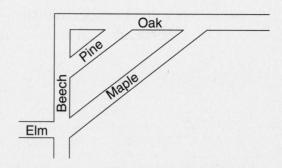

10. Which streets are perpendicular?

   a) _____ and _____

   b) _____ and _____

11. Which streets are parallel?

   a) _____ and _____

   b) _____ and _____

12. Which streets intersect with Oak?

   _____

Review Your Skills

▶ Use the protractor to find the size of each angle.

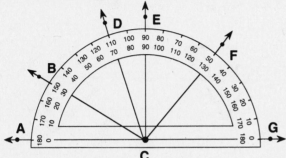

1. a) ∠ GCB _____

   b) ∠ GCF _____

   c) ∠ ACE _____

   d) ∠ ACF _____

▶ Match each angle with its measurement from the list below. Then classify the angle.
85° 45° 135° 15° 90° 165°

2. a) _____
      measurement

   b) _____
      classify

3. a) _____
      measurement

   b) _____
      classify

4. a) _____
      measurement

   b) _____
      classify

5. a) _____
      measurement

   b) _____
      classify

6. a) _____
      measurement

   b) _____
      classify

▶ Name each figure using symbols.

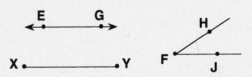

7. a) line segment YX _____

   b) line EG _____

   c) angle HFJ _____

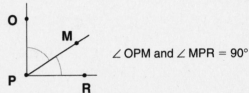

∠ OPM and ∠ MPR = 90°

8. What are these angles called?

   _____

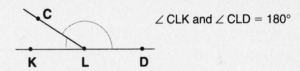

∠ CLK and ∠ CLD = 180°

9. What are these angles called?

   _____

10. If an angle measures 90°, it is called a

    _____

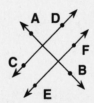

11. a) Which lines are perpendicular?

    _____ and _____

    _____ and _____

    b) Which lines are parallel?

    _____ and _____

4 POLYGONS

Drawing Polygons

Polygons are shapes made by three or more line segments that close in an area. Line segments that form the polygon are called **sides**.

▶ Count the number of sides on each traffic sign.

1.
_____
sides

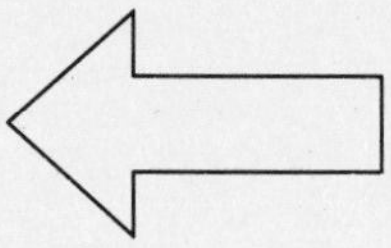

3.
_____
sides

5.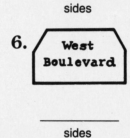
_____
sides

7.
_____
sides

2.
_____
sides

4.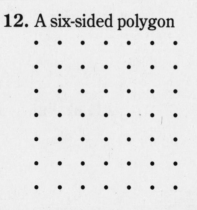
_____
sides

6.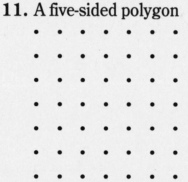
_____
sides

8.
_____
sides

▶ Connect the dots with line segments to form each of the shapes listed.

9. A three-sided polygon

11. A five-sided polygon

13. A seven-sided polygon

10. A four-sided polygon

12. A six-sided polygon

14. An eight-sided polygon

Working with Polygons

Polygons are named by the number of sides they have.

Number of Sides	Shape	Name of Polygon
3	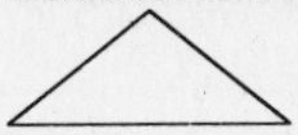 	Triangle
4		Quadrilateral
5	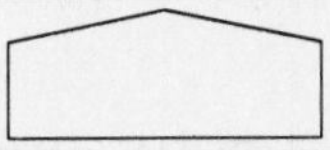	Pentagon
6		Hexagon
8		Octagon

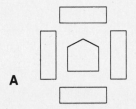

A

B

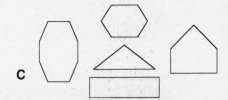

C

▶ Use the chart above to help you answer the questions.

1. Which arrangement is made up of two triangles and an octagon? _____
letter

2. Which arrangement is made up of four quadrilaterals and a pentagon? _____
letter

3. Which arrangement is made up of five different shapes? _____
letter

4. In the box below, make an arrangement with three triangles and two hexagons.

5. In the box below, make an arrangement with one octagon, two quadrilaterals, and two pentagons.

Regular Polygons

Regular polygons have sides that are the same length.

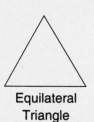
Equilateral
Triangle

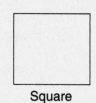
Square

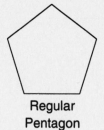
Regular
Pentagon

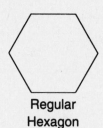
Regular
Hexagon

Regular
Octagon

▶ Draw an ✕ through the polygons that are not regular.

1. Triangle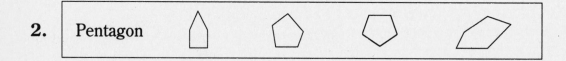

2. Pentagon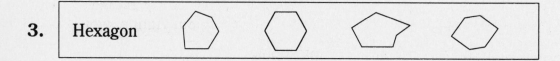

3. Hexagon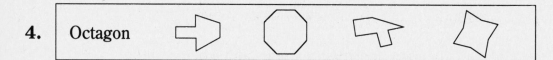

4. Octagon

▶ Write whether each polygon is regular or not.

5. _____ **6.** _____ **7.** _____ **8.** _____

Special Quadrilaterals

Remember, a four-sided polygon is called a quadrilateral. There are different kinds of quadrilaterals.

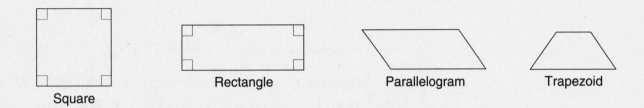

Square   Rectangle   Parallelogram   Trapezoid

▶ Name the quadrilateral described in each problem.

1. The sides are not all equal.
   It has only one pair of opposite
   sides that are parallel.

   _____

2. It has four right angles.
   Opposite sides are parallel.
   It has four equal sides.

   _____

3. Opposite sides are parallel.
   Opposite sides are equal.
   There are no right angles.

   _____

4. Only opposite sides are equal.
   Opposite sides are parallel.
   There are four right angles.

   _____

▶ Draw an ✕ through the quadrilaterals that do not belong.

5. Rectangle

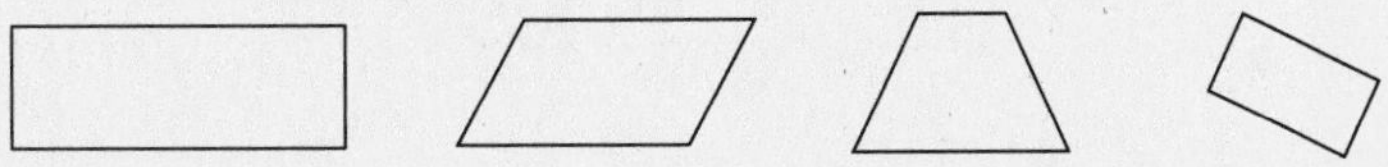

6. Parallelogram

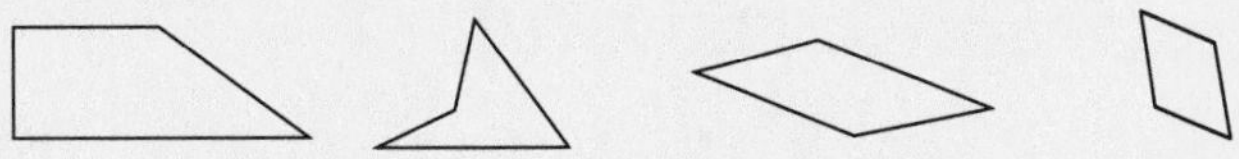

7. Trapezoid

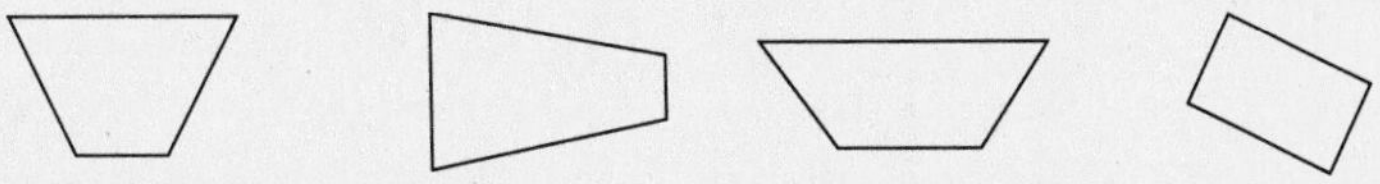

8. Square

Parallelograms

A parallelogram is a special kind of quadrilateral. Rectangles and squares are parallelograms.

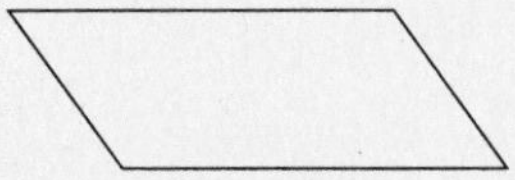

Parallelogram

Rectangle

Square

The opposite sides of a parallelogram are parallel and are the same length.

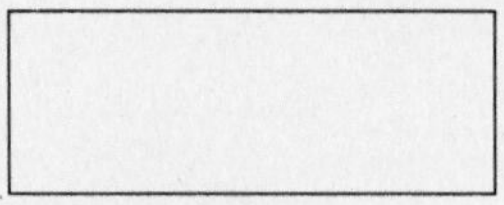

A rectangle is a parallelogram with four right angles.

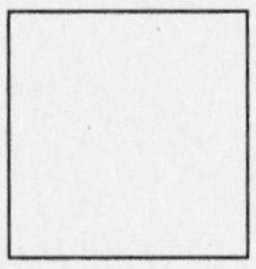

A square is a special rectangle. All four sides are the same length.

▶ Connect the dots to draw pictures of the quadrilaterals.

1. A **rectangle** but not a square

3. A **parallelogram** but not a rectangle

5. A **square**

2. A **parallelogram** with no right angles

4. A **parallelogram** with four right angles

6. A **parallelogram** with four right angles and all sides equal

5 TRIANGLES

Triangles Classified by Sides

Triangles are classified according to the proportionate lengths of their sides.

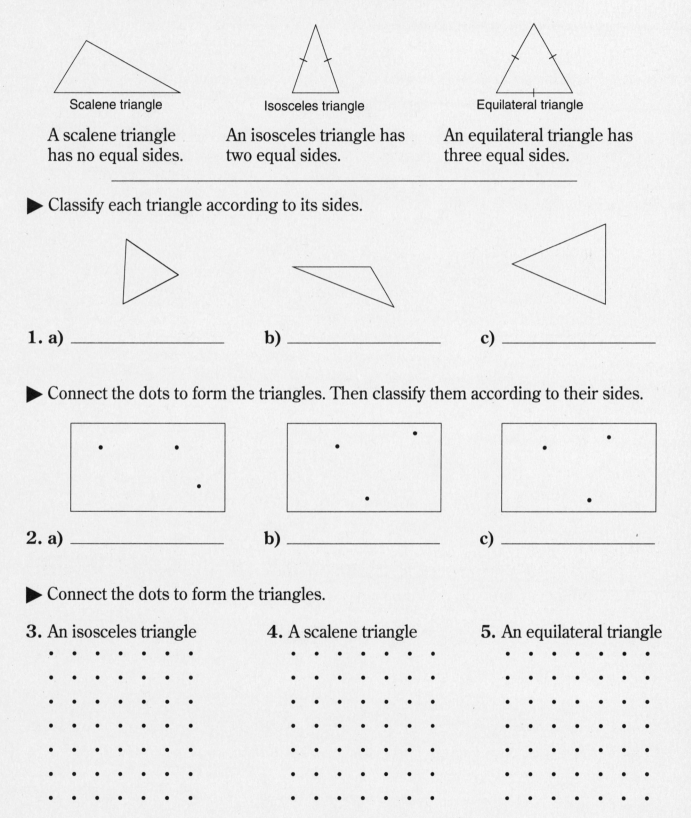

Scalene triangle

Isosceles triangle

Equilateral triangle

A scalene triangle has no equal sides.

An isosceles triangle has two equal sides.

An equilateral triangle has three equal sides.

▶ Classify each triangle according to its sides.

1. a) _____   b) _____   c) _____

▶ Connect the dots to form the triangles. Then classify them according to their sides.

2. a) _____   b) _____   c) _____

▶ Connect the dots to form the triangles.

3. An isosceles triangle   4. A scalene triangle   5. An equilateral triangle

Triangles Classified by Angles

Triangles can also be classified according to the size of their angles.

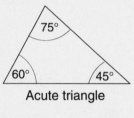

Acute triangle

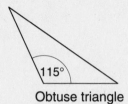

Right triangle

Obtuse triangle

An acute triangle has three acute angles.

A right triangle has a right angle.

An obtuse triangle has an obtuse angle.

▶ Classify each triangle according to its angles.

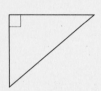 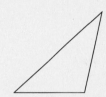 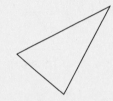 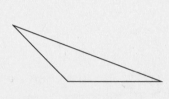

1. a) _____   b) _____   c) _____   d) _____

▶ Connect the dots to form the triangles. Then classify them according to their angles.

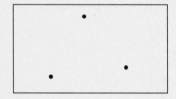 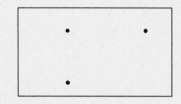 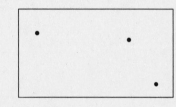

2. a) _____   b) _____   c) _____

▶ Connect the dots to form the triangles.

3. A right triangle

4. An obtuse triangle

5. An acute triangle

The Sum of Angles in a Triangle

It is important to know that the **sum** of the angles of any triangle is **180°**.

Example

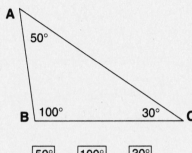

An angle in a triangle is often represented by a single letter that represents the **vertex** of the angle.

∠ BAC = ∠ A

|50°| |100°| |30°|

∠ A + ∠ B + ∠ C = 180°

▶ Find the size of the missing angle in each triangle.

Think:
125° + 30° = 155°
so 180° − 155° = _____

1. ∠ J = _____

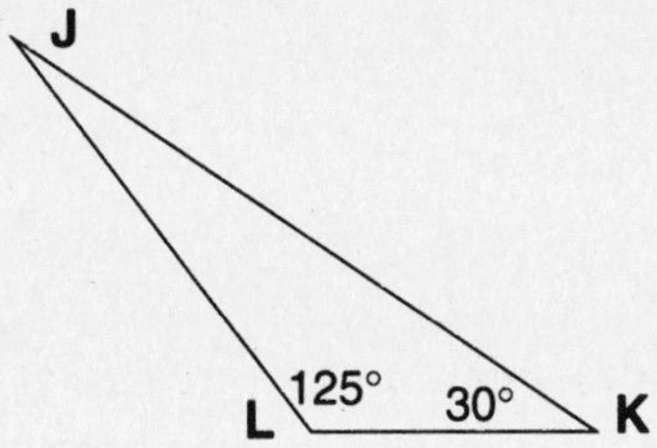

2. ∠ E = _____

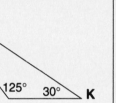

3. ∠ N = _____

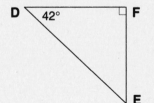

4. ∠ T = _____

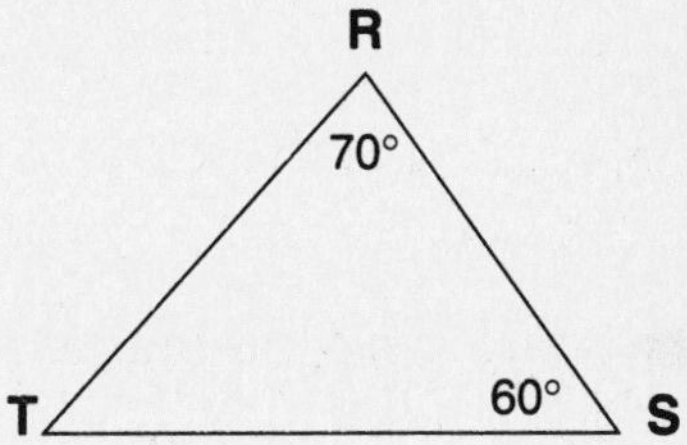

5. ∠ D = _____

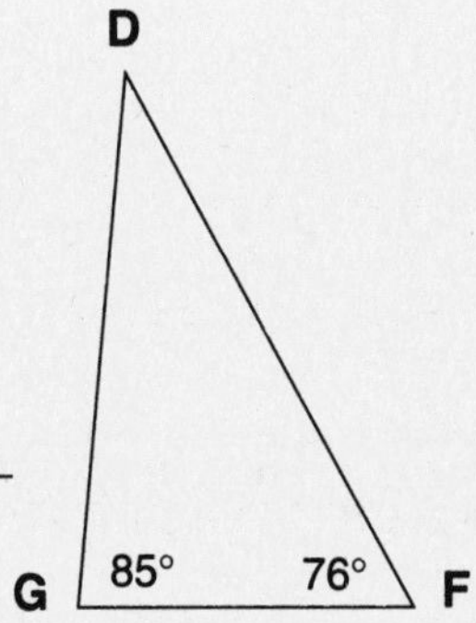

6. ∠ B = _____

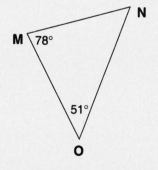

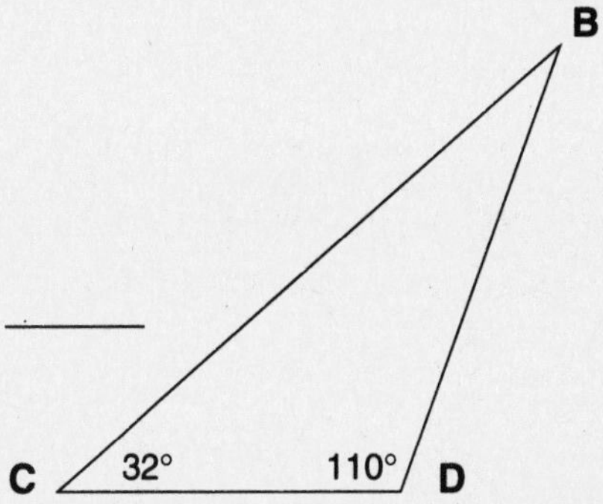

Review of Polygons and Triangles

▶ Match the name of the polygon with the correct figure. There may be more than one figure that matches each name. Figures may be used more than once.

1. Parallelograms ___B,D,G___
2. Squares _____
3. Rectangles _____
4. Pentagons _____
5. Quadrilaterals _____
6. Hexagons _____
7. Regular Polygons _____
8. Trapezoids _____
9. Octagons _____

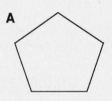 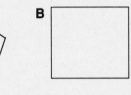 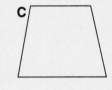 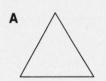 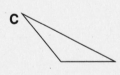  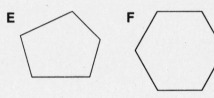

▶ Write the letter that matches each triangle.

10. Scalene, obtuse _____
11. Isosceles, right _____
12. Equilateral _____
13. Scalene, acute _____

▶ Find the size of the missing angle.

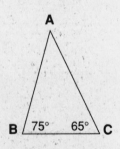 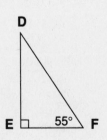

14. a) ∠ A = _____   b) ∠ H = _____   c) ∠ D = _____

Tiling Floors

▶ When tiling floors, it is important to cut tiles to fit the corners. The shaded area shows the part that is cut and not used for each tile. For each tile, write the letter that matches the corner angle.

Corner Angles

1. _____

3. _____

2. _____

4. _____

A

C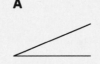

B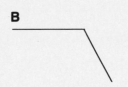

D

▶ The tiles below have already been cut. For each tile, write the letter that matches the corner angle.

Corner Angles

5. _____

7. _____

6. _____

8. _____

A

C

B

D

▶ Draw lines on the tiles to show where to cut for each corner angle. Shade the part that is not used for each tile.

9.

10.

11.

12.

Measure of Length

To measure the length of something, you must select a unit of measure and then count how many units there are in that length.

▶ Measure the length of each figure using the given unit of measure. The first one is done for you.

Measure	Unit	Figure

1. __5__

2. _____

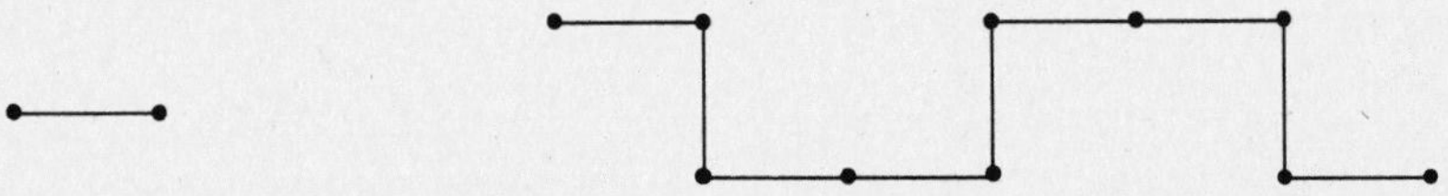

3. _____

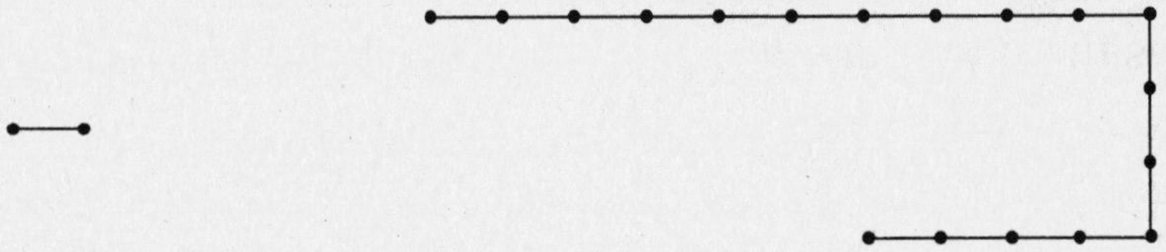

4. _____

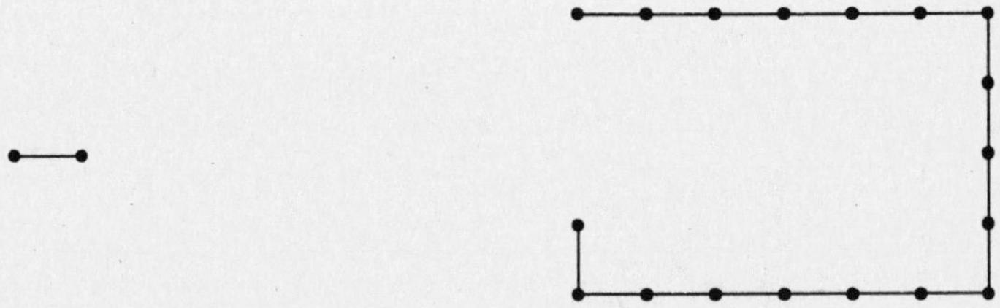

5. _____

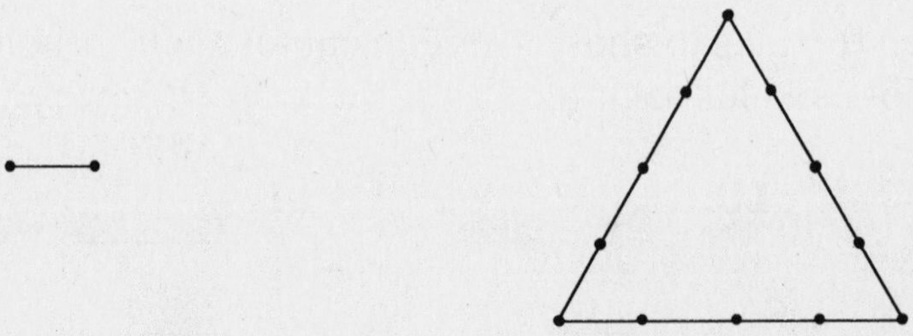

Distance Around

The **perimeter** of a figure is the length of the distance completely around the outside of the shape.

Example

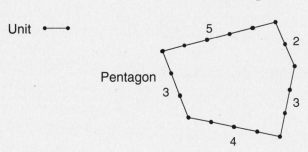

$$5 + 2 + 3 + 4 + 3 = 17$$

Perimeter

The perimeter of the pentagon is 17 units.

▶ Find the perimeter for each figure.

•——• is one unit

1.

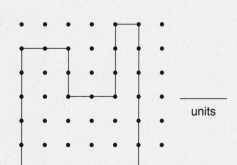

_____ units

2.

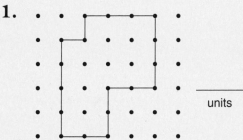

_____ units

3.

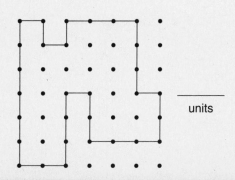

_____ units

4. Draw a figure that has a perimeter of 20 units.

5. Draw a figure that has a perimeter of 28 units.

6. Draw a figure that has a perimeter of 18 units.

Find the Perimeter

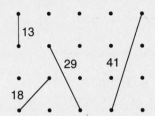

▶ Use the lengths above to find the perimeter of each polygon.

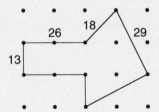

1. _____

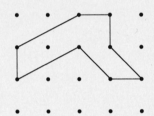

4. _____

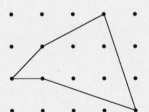

2. _____

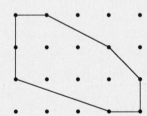

5. _____

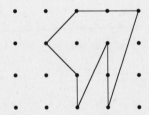

3. _____

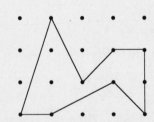

6. _____

Missing Lengths of Polygons

There are times when the lengths of sides of polygons are not given.

▶ Find and label the missing lengths of the sides below. (All angles are right angles.)

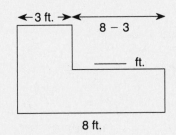

1. The missing length is _____ feet.

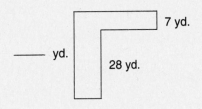

3. The missing length is _____ yards.

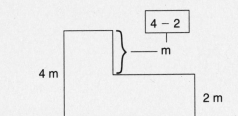

2. The missing length is _____ meters.

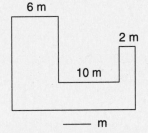

4. The missing length is _____ meters.

▶ Find the perimeter of each polygon.

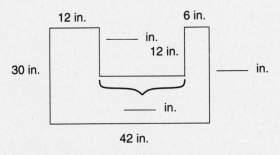

5. The perimeter is _____ inches.

7. The perimeter is _____ centimeters.

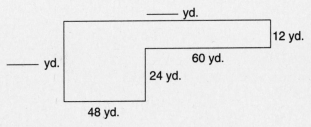

6. The perimeter is _____ feet.

8. The perimeter is _____ yards.

Perimeter Formulas

A formula is a short way of stating a rule. You can use formulas to find perimeters.

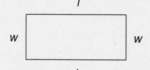

Square

$$P = 4s$$

The perimeter (P) of a square is equal to four times one side (s).

Rectangle
(Length is the longer side.)

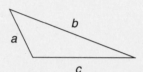

$$P = 2l + 2w$$

The perimeter (P) of a rectangle is equal to twice the length (l) plus twice the width (w).

Triangle

$$P = a + b + c$$

The perimeter (P) of a triangle is equal to the sum of its sides (a, b, c).

▶ Use the formulas above to find the perimeters.

7 in.

7

1. a) ___$P = 4s$___
formula

b) ___$P = 4 \times 7$___
substitute known values

c) $P =$ _____ inches
solution

8 ft.

7 ft.   9 ft.

3. a) _____
formula

b) _____
substitute known values

c) $P =$ _____ feet
solution

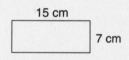

15 cm

7 cm

5. a) _____
formula

b) _____
substitute known values

c) $P =$ _____ centimeters
solution

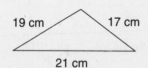

19 cm   17 cm

21 cm

2. a) _____
formula

b) _____
substitute known values

c) $P =$ _____ centimeters
solution

12 m

5 m

4. a) _____
formula

b) _____
substitute known values

c) $P =$ _____ meters
solution

19 yd.

6. a) _____
formula

b) _____
substitute known values

c) $P =$ _____ yards
solution

Perimeter Applications

▶ Use the drawings to help answer the questions. Label all answers.

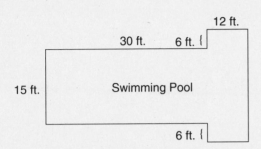

1. What is the perimeter of the swimming pool? _____

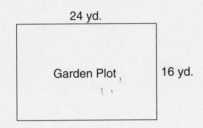

2. How many yards of fencing will be needed to fence in the garden plot? _____

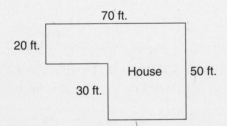

3. What is the perimeter of the house? _____

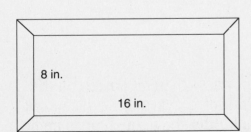

4. What is the perimeter of the inside of the picture frame? _____

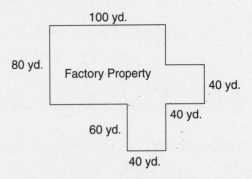

5. A factory wants to build a fence around its property. How many yards of fence will be needed? _____

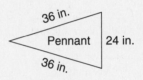

6. How much braid will be needed to go around the pennant? _____

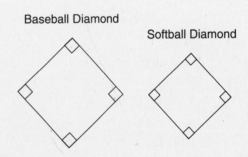

7. There are 90 feet between bases on a baseball diamond and 60 feet between bases on a softball diamond. How many feet shorter is the distance around the softball diamond? _____

Applying Perimeter Skills

Remember, formulas are a great way of organizing information.

▶ Use the following perimeter formulas to solve the problems. Drawing a picture can also help you organize your thoughts.

Square: $P = 4s$	Rectangle: $P = 2l + 2w$	Triangle: $P = a + b + c$

1. Mr. Perez wants to fence in his garden. His garden is shaped like a rectangle and is 28 feet long and 18 feet wide. What length fence does he need?

   a) _____

formula

   b) _____

substitute known values

   c) _____

solution

2. A square parking lot has sides of 85 yards each. What is the perimeter of the parking lot?

   a) _____

formula

   b) _____

substitute known values

   c) _____

solution

3. A rectangular picture has a length of 25 inches and a width of 12 inches. What is the perimeter of the picture?

   a) _____

formula

   b) _____

substitute known values

   c) _____

solution

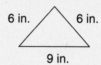

6 in.   6 in.

9 in.

4. Janet made decorations shaped like triangles. If the sides measure 6 inches, 9 inches, and 6 inches, how much ribbon does she need to trim the edges of each decoration? _____

_____ yd.

_____ yd.

5. Nancy wants to fence in a square lot. Each side of the lot measures 35 yards. How many yards of fence are left over if she bought 150 yards of fencing?

   _____

_____ ft.

_____ ft.

_____ ft.

6. Alan wants to plant flowers in the corner of his yard. If he uses wood as a border, how much does he need for a border that is 5 feet by 5 feet by 7 feet? _____

Surface Inside a Figure

To measure the amount of surface inside a figure, you must select a unit of measure and then count how many units it takes to exactly cover the shape.

▶ Find the amount of surface inside each figure using the given unit of measure.

Unit

1.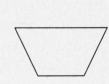

a) __2__   b) ____   c) ____

2.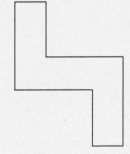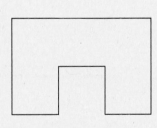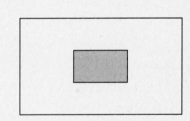

a) ____   b) ____   c) ____

3.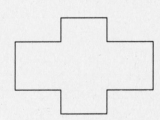

a) ____   b) ____   c) ____

4.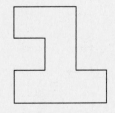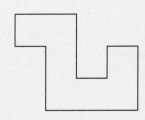

a) ____   b) ____   c) ____

Square Units

Area is a measure of surface. The area of a figure can be measured by the number of square units that are needed to exactly cover the shape of the figure.

Example

1 square unit

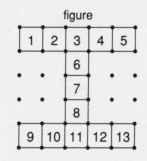
figure

Thirteen square units fit inside the figure. The area of the figure is thirteen square units.

▶ Find the area of each figure by counting the square units that are needed to exactly cover the shape. It may be helpful to connect the dots to form square units.

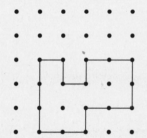

1. _____ square units

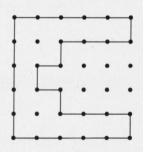

4. _____ square units

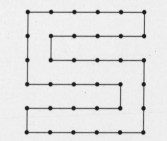

7. _____ square units

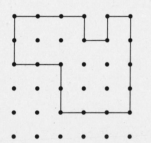

2. _____ square units

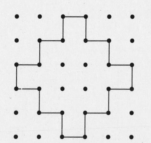

5. _____ square units

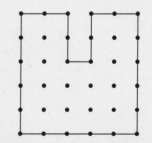

8. _____ square units

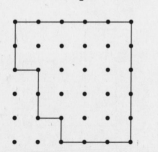

3. _____ square units

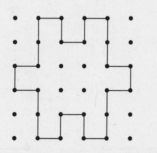

6. _____ square units

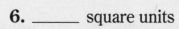

9. _____ square units

Square Inches and Square Centimeters

Square inches and square centimeters are standard units used to find area.

1 square inch

Each side of the square is one inch long.

1 square centimeter

Each side of the square is one centimeter long.

▶ Find the area of each figure below using one square inch as the unit of measure.

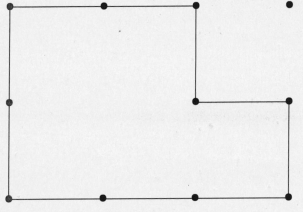

1. _____ square inches

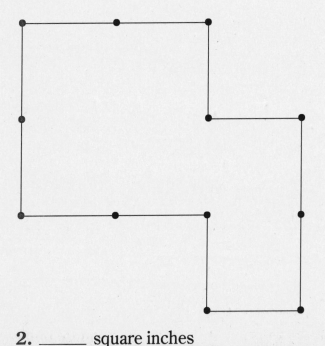

2. _____ square inches

▶ Find the area of each figure below using one square centimeter as the unit of measure.

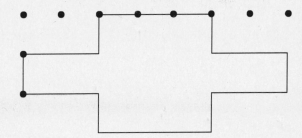

3. _____ square centimeters

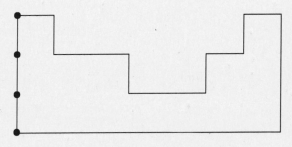

4. _____ square centimeters

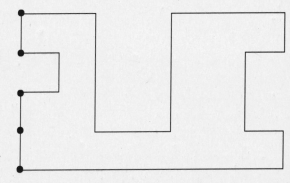

5. _____ square centimeters

Rectangles

To find the area (*A*) of a rectangle, multiply the length (*l*) by the width (*w*).

$$A = lw$$ lw means the same as $l \times w$

Example A

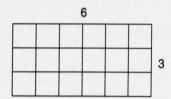

6

3

Number of rows: 3
Square units in each row: 6
3 × 6 = 18
Area = 18 square units

Example B

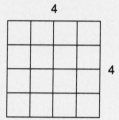

4

4

Number of rows: 4
Square units in each row: 4
4 × 4 = 16
Area = 16 square units

Example C

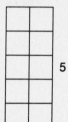

2

5

Number of rows: 5
Square units in each row: 2
5 × 2 = 10
Area = 10 square units

▶ Find the area of each rectangle.

4 ft.

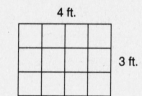

3 ft.

1. __4__ × __3__ = _____

_____ square feet

36 in.

18 in.

2. __36__ × __18__ = _____

_____ square inches

17 cm

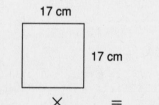

17 cm

3. _____ × _____ = _____

_____ square centimeters

80 yd.

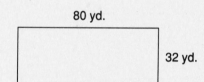

32 yd.

4. _____ square yards

48 m

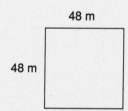

48 m

5. _____ square meters

65 cm

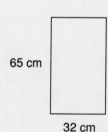

32 cm

6. _____ square centimeters

Parallelograms

The area of a parallelogram can be discovered by finding the area of the rectangle hidden within the original shape. Picture the parallelogram as a rectangle.

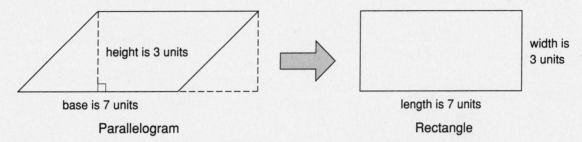

The area of both the rectangle and the parallelogram is 21 square units.

$$A = bh$$

bh means the same as $b \times h$

$$21 = 7 \times 3$$

To find the area of a parallelogram, multiply the base by the height.

▶ Find the area of each parallelogram.

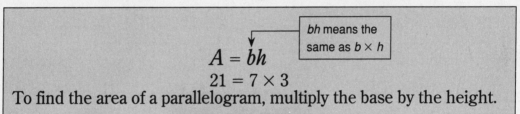

1. _____ square inches

4. _____ square meters

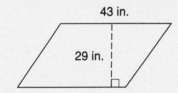

2. _____ square centimeters

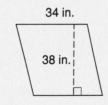

5. _____ square inches

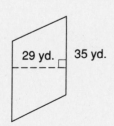

3. _____ square yards

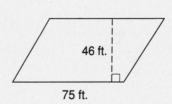

6. _____ square feet

Triangles

The area of a triangle can be discovered by using the area of the parallelogram. Any parallelogram can be divided into two triangles of the same shape and size.

Example

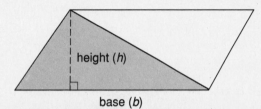

height (*h*)

base (*b*)

Area of the parallelogram: $b \times h$

Area of the triangle: $\frac{1}{2}$ of ($b \times h$)

$$A = \tfrac{1}{2}bh$$

The area (*A*) of the triangle is equal to one-half times the base (*b*) times the height (*h*).

▶ Find the area of each triangle.

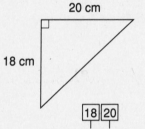

20 cm

18 cm

18 20

1. a) $A = \frac{1}{2}\,b\,h$

  formula

  b) $A = \frac{1}{2} \times 18 \times 20$

  substitute known values

  c) $A =$ _____ square centimeters

  solution

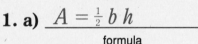

22 yd.

55 yd.

2. a) _____

  formula

  b) _____

  substitute known values

  c) $A =$ _____ square yards

  solution

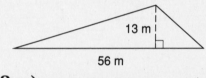

13 m

56 m

3. a) _____

  formula

  b) _____

  substitute known values

  c) $A =$ _____ square meters

  solution

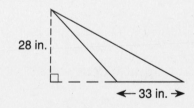

28 in.

← 33 in. →

4. a) _____

  formula

  b) _____

  substitute known values

  c) $A =$ _____ square inches

  solution

Practice Your Skills

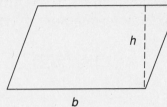

$$A = lw$$

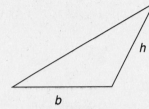

$$A = bh$$

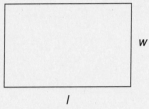

$$A = \tfrac{1}{2} bh$$

► Choose the correct formula. Then find the area of each polygon. Correctly label your answers in square units.

1.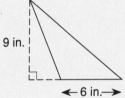
9 in.
← 6 in. →

a) _____
formula

b) _____
area

5.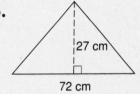
27 cm
72 cm

a) _____
formula

b) _____
area

2. 25 yd.
25 yd.

a) _____
formula

b) _____
area

6.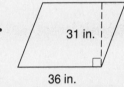
31 in.
36 in.

a) _____
formula

b) _____
area

3.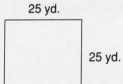
12 m
← 18 m →

a) _____
formula

b) _____
area

7.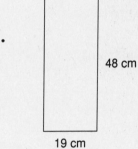
48 cm
19 cm

a) _____
formula

b) _____
area

8. 75 yd.
75 yd.

a) _____
formula

b) _____
area

4.
26 in.
20 in.

a) _____
formula

b) _____
area

Dividing Figures into Parts

When finding the area of a polygon, it is often necessary to divide the figure into parts that form rectangles or triangles.

Example

Step 1

Find the area of the polygon.

Divide the figures into two parts: one rectangle and one triangle.

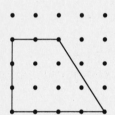

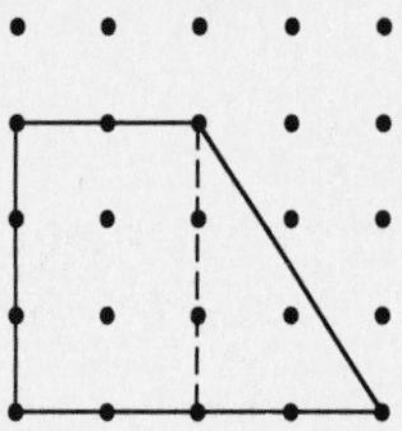

Step 2

Find the area of the rectangle: 6
Find the area of the triangle: 3

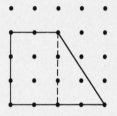

The area of the polygon is 9 square units.

▶ Use a dotted line to divide each figure into parts that form rectangles or triangles.

1.

2.

3.

4.

▶ Find the area of each figure. Use dotted lines to divide each figure into parts that form rectangles or triangles. Correctly label your answers.

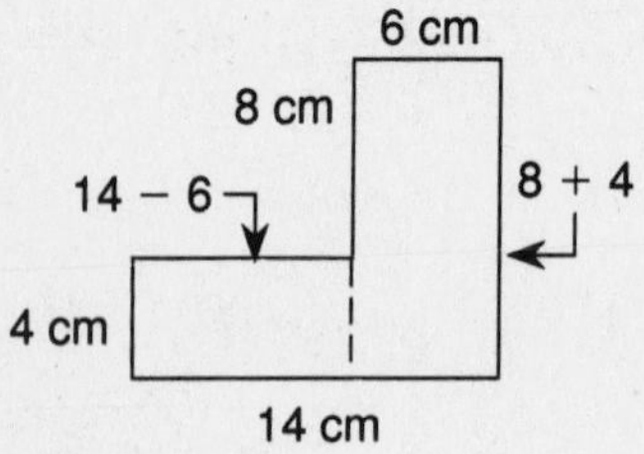

6 cm
8 cm
14 − 6
4 cm
8 + 4
14 cm

5. _____

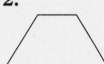

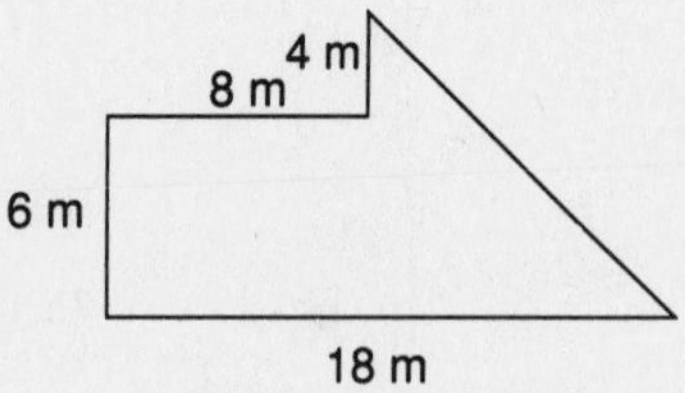

4 m
8 m
6 m
18 m

7. _____

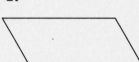

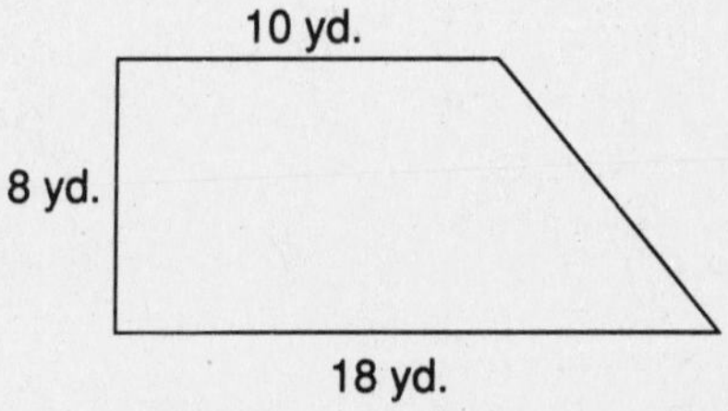

10 yd.
8 yd.
18 yd.

9. _____

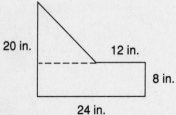

20 in.
12 in.
8 in.
24 in.

6. _____

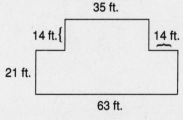

35 ft.
14 ft.{
14 ft.
21 ft.
63 ft.

8. _____

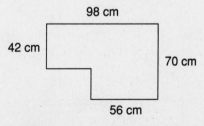

98 cm
42 cm
70 cm
56 cm

10. _____

Drawing to Find Areas and Perimeters

▶ Connect the dots to draw figures with the following **areas.**

1. 10 square units

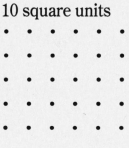

2. $21\frac{1}{2}$ square units

3. $8\frac{1}{2}$ square units

▶ Connect the dots to draw figures with the following **perimeters.** Use only horizontal and vertical line segments.

4. 14 units

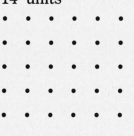

5. 22 units

6. 18 units

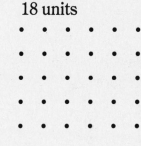

▶ Draw a figure for each **perimeter** and **area.** Use only horizontal and vertical line segments to connect the dots.

7.

Perimeter=14 Area=10

8.

Perimeter=18 Area=20

9.

Perimeter=16 Area=15

Perimeter and Area Review

1. Use letters to arrange the polygons from smallest to largest perimeter.

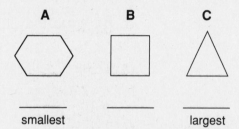

A B C

_____   _____   _____
smallest   largest

2. Use letters to arrange the polygons from smallest to largest area.

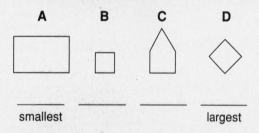

A B C D

_____   _____   _____   _____
smallest   largest

3. How many tiles are needed to cover the floor? _____

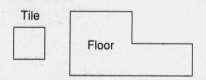

4. Find the perimeter of each figure. Correctly label your answers.

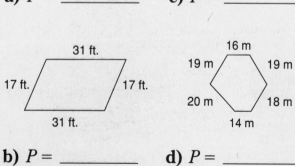

a) $P =$ _____   **c)** $P =$ _____

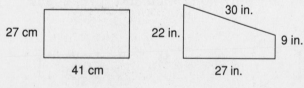

b) $P =$ _____   **d)** $P =$ _____

5. Find the area of each figure. Correctly label your answers.

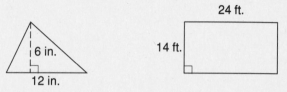

a) $A =$ _____   **c)** $A =$ _____

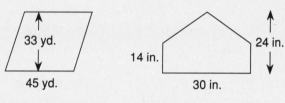

b) $A =$ _____   **d)** $A =$ _____

6. Find the perimeter and area of each figure.

a) 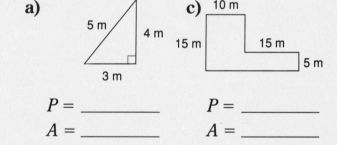 **c)**

$P =$ _____ $P =$ _____

$A =$ _____ $A =$ _____

b) 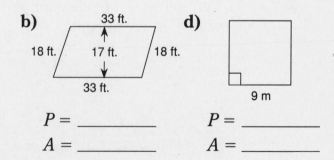 **d)**

$P =$ _____ $P =$ _____

$A =$ _____ $A =$ _____

Using Perimeter and Area

▶ Take a minute to think about whether you are to find the perimeter or area. Then label and solve each problem.

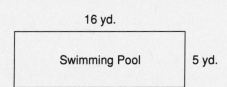

16 yd.

Swimming Pool   5 yd.

1. How many yards of nonskid tape will be needed to go around the pool?

a) _____
      perimeter or area

b) _____
      answer

2. How many square yards of material will be needed to cover the pool?

a) _____
      perimeter or area

b) _____
      answer

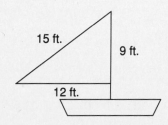

15 ft.   9 ft.

12 ft.

3. How many square feet of material will be needed to make the sail?

a) _____
      perimeter or area

b) _____
      answer

4. How many feet of border are needed to trim the edge of the sail?

a) _____
      perimeter or area

b) _____
      answer

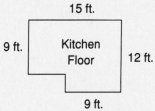

15 ft.

9 ft.   Kitchen Floor   12 ft.

9 ft.

5. How many square floor tiles (measuring 1 foot per side) will it take to cover the kitchen floor?

a) _____
      perimeter or area

b) _____
      answer

6. How many feet of baseboard molding will it take to go around the kitchen floor?

a) _____
      perimeter or area

b) _____
      answer

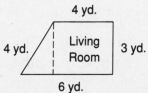

4 yd.

4 yd.   Living Room   3 yd.

6 yd.

7. How much carpeting is needed to cover the living room floor?

a) _____
      perimeter or area

b) _____
      answer

8. How many yards of tackboard are needed to go around the edge of the living room?

a) _____
      perimeter or area

b) _____
      answer

Perimeter and Area Applications

▶ Use the drawings to help answer the questions.

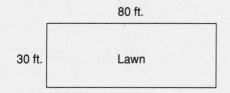

80 ft.

30 ft.   Lawn

1. a) How much grass seed will the lawn need if a pound of seed covers 800 square feet? _____

    b) At $2.45 per pound, how much will the grass seed cost? _____

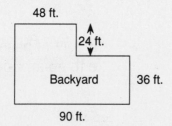

48 ft.

24 ft.

Backyard   36 ft.

90 ft.

2. a) Daniel wants to fertilize his backyard. Fertilizer is sold in bags that cover 5,000 square feet. Will 1 bag be enough? _____

    b) How many feet of fencing will it take to go around the backyard? _____

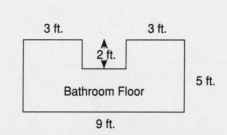

3 ft.   3 ft.

2 ft.

Bathroom Floor   5 ft.

9 ft.

3. a) Each square foot of tile costs $3.95. How much will it cost to tile the bathroom floor? _____

    b) How many feet of baseboard molding will it take to go around the bathroom floor? _____

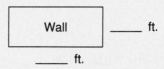

Backyard _____ ft.

_____ ft.

4. The Coys want to fence in their backyard, which is 35 feet wide and 125 feet long. How much fencing do they need? _____

Wall _____ ft.

_____ ft.

5. A gallon of paint will cover 450 square feet. Will this be enough to cover 4 walls 15 feet long and 8 feet high?

_____

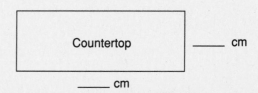

Countertop _____ cm

_____ cm

6. A carpenter wants to put a metal strip around a countertop that is 150 centimeters long and 48 centimeters wide. How much stripping will be needed? _____

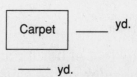

Carpet _____ yd.

_____ yd.

7. Gloria wants to carpet a room that is 4 yards wide and 6 yards long. If the carpet costs $15.95 a square yard, how much will it cost? _____

Reading House Plans

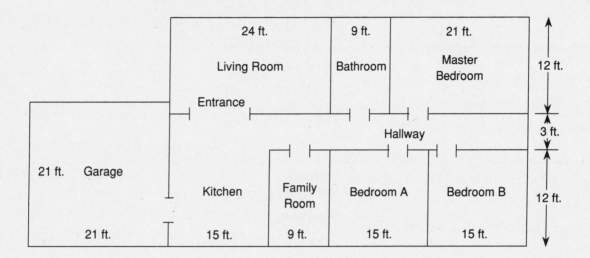

▶ The diagram shows the floor plan for a new house. Each doorway is three feet wide. Use the floor plan to answer each question.

1. How much baseboard will be needed to go around each room? Subtract 3 feet for each doorway and 9 feet for the living room entrance.

   a) Living room _____   **d)** Master bedroom _____

   b) Family room _____   **e)** Bedroom A _____

   c) Bathroom _____   **f)** Bedroom B _____

2. If baseboard costs $.55 a foot, how much will it cost for the 6 rooms? _____

3. How much will it cost to cement the garage floor at $1.85 a square foot? _____

> 9 square feet = 1 square yard

4. How much will it cost to install new carpet in the living room at $27.50 a square yard? _____

5. How much will it cost to install new carpet in each of the bedrooms at $21.50 a square yard?

   a) Master bedroom _____

   b) Bedroom A _____

   c) Bedroom B _____

8 CIRCLES

Learning about Circles

Circle

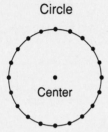

Center

Radius

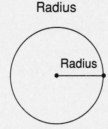

Radius

Diameter

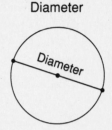

Diameter

A **circle** is a closed curve whose points are all the same distance from the center.

The **radius** is a line segment from the center to any point on the circle.

The **diameter** is a line segment connecting two points on the circle that passes through the center.

The **diameter** (d) of a circle is equal to 2 times the radius (r): $d = 2r$.

The **radius** (r) of a circle is equal to the diameter (d) divided by 2: $r = \frac{d}{2}$.

▶ Use the formulas above to answer the following questions.

1. What is the diameter of each circle?

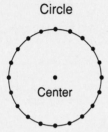 28 ft. $6\frac{1}{2}$ in.   15 cm   3 m

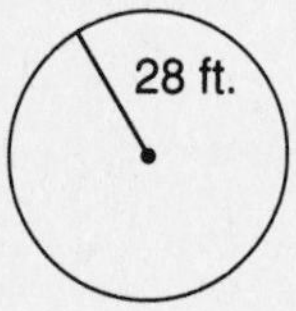

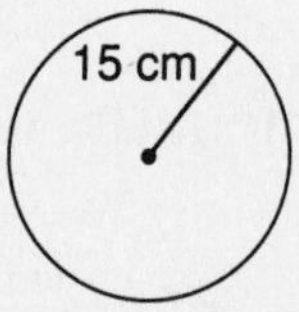

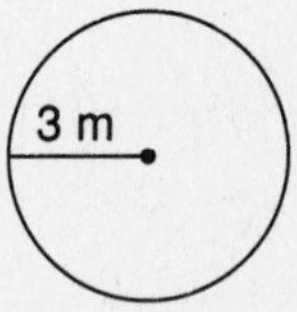

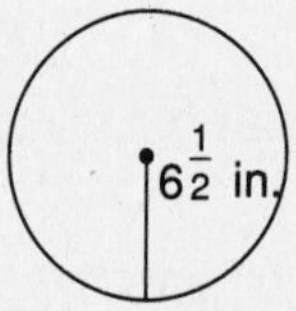

a) _____   b) _____   c) _____   d) _____

2. What is the radius of each circle?

 16 in.   30 cm   8 yd.   44 m

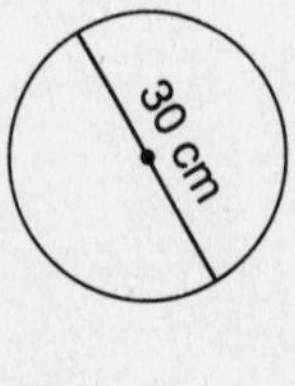

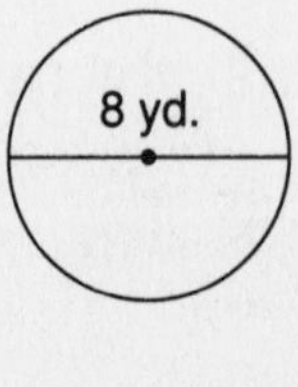

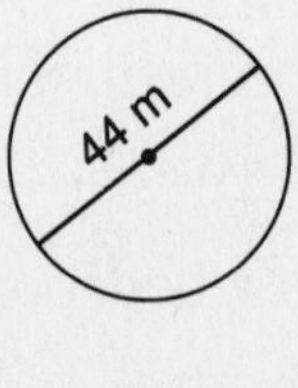

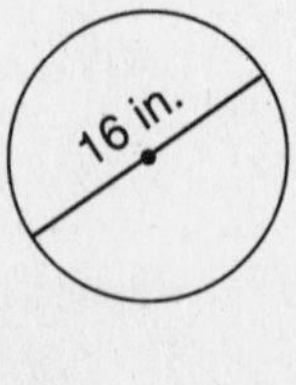

a) _____   b) _____   c) _____   d) _____

Finding the Circumference

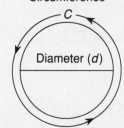

Circumference

C

Diameter (*d*)

The distance around a circle is called its **circumference** (*C*).

For any circle, no matter what its size, $C \div d$ is 3.14 when rounded to the nearest hundredth. The Greek letter π (pi) is used for this ratio.

> The **circumference** (*C*) of a circle is equal to π (pi) times the diameter (*d*).
> $$C = \pi d$$

▶ Find the circumference of each circle. Use 3.14 as π. Label your answers.

1. Diameter = 5 feet

a) $C = \pi d$ 3.14   5
     formula

b) $C = 3.14 \times 5$
     substitute known values

c) $C = $ _____ feet
     solution

2. Diameter = 25 inches

a) _____
     formula

b) _____
     substitute known values

c) $C = $ _____
     solution

3. Diameter = 9.5 meters

a) _____
     formula

b) _____
     substitute known values

c) $C = $ _____
     solution

4. Diameter = 14 centimeters

a) _____
     formula

b) _____
     substitute known values

c) $C = $ _____
     solution

5. Radius = 6 centimeters

a) $C = \pi d$ $d = 2r$
     formula

b) $C = $ ___ $\times 12$
     substitute known values

c) $C = $ _____ centimeters
     solution

6. Radius = 3 yards

a) _____
     formula

b) _____
     substitute known values

c) $C = $ _____
     solution

7. Radius = 25 centimeters

a) _____
     formula

b) _____
     substitute known values

c) $C = $ _____
     solution

8. Radius = 4.5 inches

a) _____
     formula

b) _____
     substitute known values

c) $C = $ _____
     solution

Area of a Circle

> The **area** of a circle is equal to π (pi) times the radius (r) squared.
> $$A = \pi r^2$$

Example A

Find the area of a circle with a radius of 8 centimeters.

r^2 means to multiply the radius by itself

$$\boxed{r \times r}$$
$$A = \pi r^2$$
$$A = 3.14 \times 8 \times 8$$
$$A = 200.96 \text{ square centimeters}$$

Example B

Find the area of a circle with a diameter of 4 inches.

Step 1: Find the radius. It is one-half the diameter or 2 inches.
Step 2: Use the formula.

$$A = \pi r^2$$
$$A = 3.14 \times 2 \times 2$$
$$A = 12.56 \text{ square inches}$$

▶ Find the area of each circle. Use 3.14 as π.

1. Radius = 5 centimeters

   a) _____
          formula

   b) _____
     substitute known values

   c) $A =$ _____ square centimeters
         solution

2. Radius = 12 inches

   a) _____
          formula

   b) _____
     substitute known values

   c) $A =$ _____ square inches
         solution

3. Radius = 65 meters

   a) _____
          formula

   b) _____
     substitute known values

   c) $A =$ _____ square meters
         solution

4. Diameter = 8 millimeters

   a) _____
          formula

   b) _____
     substitute known values

   c) $A =$ _____ square millimeters
         solution

5. Diameter = 22 centimeters

   a) _____
          formula

   b) _____
     substitute known values

   c) $A =$ _____ square centimeters
         solution

6. Diameter = 40 inches

   a) _____
          formula

   b) _____
     substitute known values

   c) $A =$ _____ square inches
         solution

Using Circles

▶ Read each problem carefully. Decide if the problem asks for the area or the circumference of the circle, then solve. Drawing a picture may help you decide. Use 3.14 for π when needed.

1. A sprinkler waters in a circular pattern. It sprays water out a distance of 20 feet. To the nearest square foot, how much lawn surface does the sprinkler cover? _____

2. A round swimming pool has a radius of 8 feet. To the nearest foot, what is the pool's circumference? _____

3. To the nearest square meter, what is the area of a circular rose garden that has a radius of 5 meters? _____

4. A circular garden has a diameter of 25 feet. To the nearest foot, how many feet of fencing material will be needed to fence in the garden? _____

5. To the nearest square meter, what is the surface area of a round fish pond 15 meters in diameter? _____

6. How much trim is needed for a round tablecloth 6 feet in diameter? _____

7. What is the area of the shaded walkway around the garden? _____ square meters

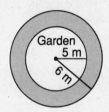

8. How many square centimeters are not used for the can lid? _____

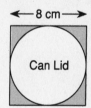

9. What is the area of the shaded area? _____ square millimeters

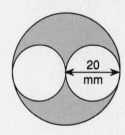

10. What is the area of the shaded area? _____ square centimeters

Count the Cubes

A solid figure takes up space and has volume. **Volume** is the number of cubic units (cubes) needed to fill a space. One way to find the volume is to count the cubes.

▶ Count the number of cubes in each solid figure.

Cubic Unit	Solid	Volume

All edges of the cubic unit are the same length.

1. 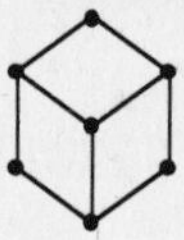 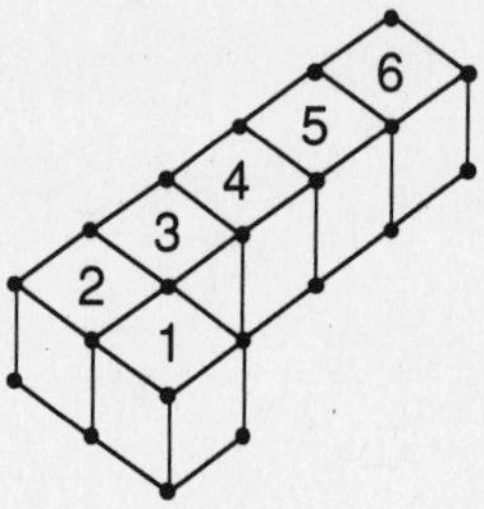 _____ cubes

2. 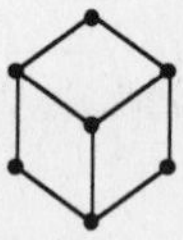 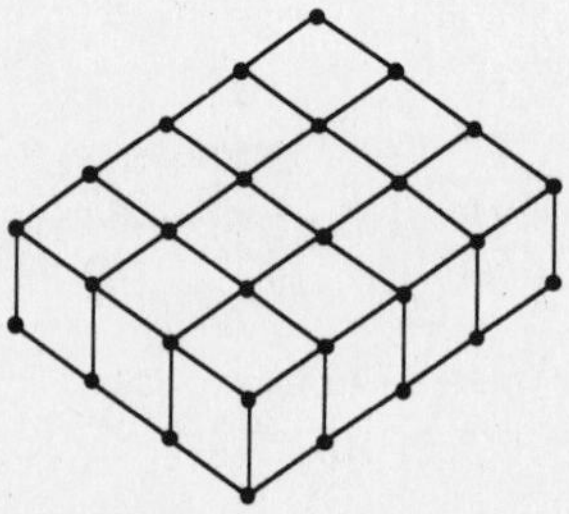 _____ cubes

3. 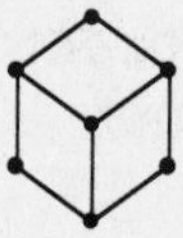 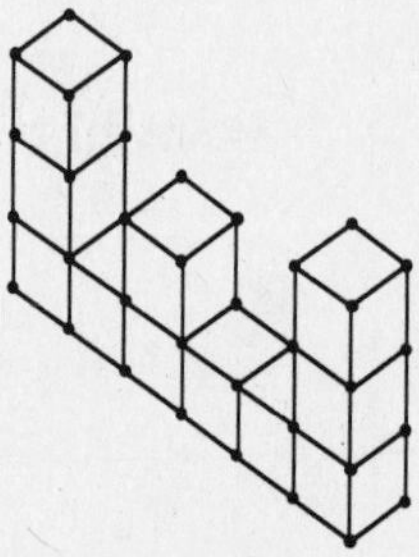 _____ cubes

4. 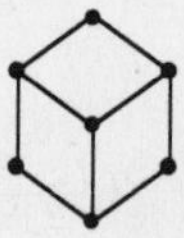  _____ cubes

5. 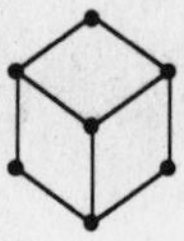  _____ cubes

Relating a Solid to a Drawing

The volume of a solid figure is the number of cubic units that it contains.

▶ For each solid, copy the drawing that is shown; then find the volume by counting the cubes. It may be helpful to connect the dots to form cubes.

1.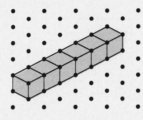

_____ cubic units

5.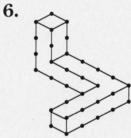

_____ cubic units

2.

_____ cubic units

6.

_____ cubic units

3.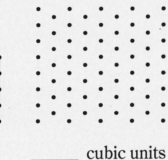

——— cubic units

7.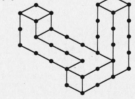

_____ cubic units

4.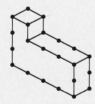

_____ cubic units

8.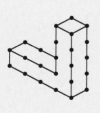

_____ cubic units

Hidden Cubes

Sometimes cubes can be hidden by other cubes. Don't forget to count the cubes you cannot see.

Example

Solid

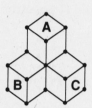

The volume of the solid is 4 cubic units. Cube D is directly below cube A.

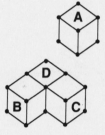

▶ For each solid, copy the drawing that is shown. Then find the volume by counting the cubes.

1.

_____ cubic units

4.

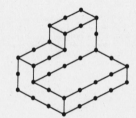

_____ cubic units

2.

_____ cubic units

5.

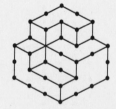

_____ cubic units

3.

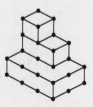

_____ cubic units

6.

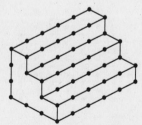

_____ cubic units

Finding Volume

Example

| Find the volume of the rectangular solid. | Step 1: Multiply 3 × 6 to find the number of cubic units on the bottom layer. | Step 2: There are 4 layers of 18 cubes. 4 × 18 = 72 The volume of the rectangular solid is 72 cubic units. |

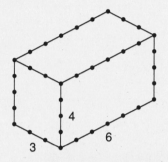

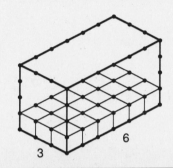

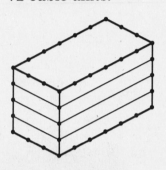

> To find the **volume** (V) of a rectangular solid, multiply the length (l) by the width (w) by the height (h).
>
> $$V = lwh$$

▶ Find the volume of each rectangular solid.

1.

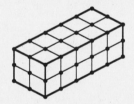

a) $V = $ ___ × ___ × ___
b) $V = $ ___ cubic units

3.

a) $V = $ ___ × ___ × ___
b) $V = $ ___ cubic units

2.

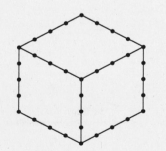

a) $V = $ ___ × ___ × ___
b) $V = $ ___ cubic units

4.

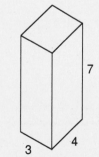

a) $V = $ ___ × ___ × ___
b) $V = $ ___ cubic units

Apply Your Skills

▶ Find the volume of each rectangular solid. Correctly label all answers.

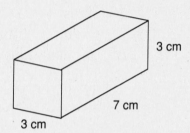

1. _____ cubic centimeters

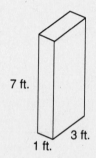

2. _____

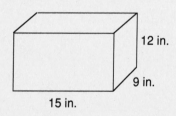

3. _____

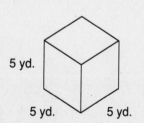

4. _____

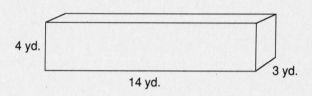

5. _____

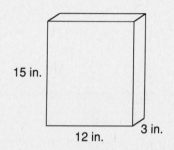

6. _____

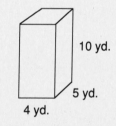

7. _____

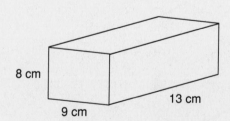

8. _____

Two-Step Volume Problems

Sometimes it is necessary to separate a solid into two or more parts to find the volume.

Example: Find the volume of the solid.

Separate the solid into two parts.

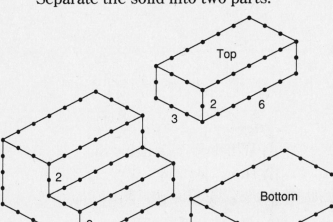

Step 1: Find the volume of the top.
6 × 3 × 2 = 36 cubic units

Step 2: Find the volume of the bottom.
6 × 5 × 2 = 60 cubic units

Step 3: Add the volume of the top and bottom. 36 + 60 = 96

The volume of the figure is 96 cubic units.

▶ Find the volume of each solid.

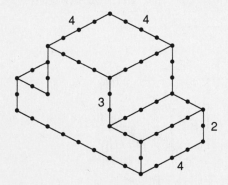

1. _____ cubic units

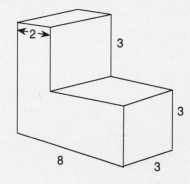

3. _____ cubic units

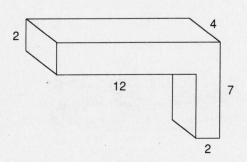

2. _____ cubic units

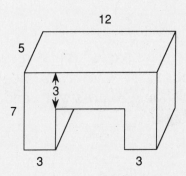

4. _____ cubic units

Volume Applications

▶ Solve the following problems. Drawing a picture may help organize the information.

1. A pool is 15 feet long, 10 feet wide, and 5 feet deep. How many cubic feet of water are needed to fill the pool? _____

2. A storage bin measures 9 meters by 5 meters by 8 meters. How many cubic meters does the bin enclose?

_____

3. A sandbox is 6 feet long, 3 feet wide, and 2 feet high. What is the volume of the sandbox? _____

4. What is the volume of a freight car 20 meters long, 10 meters wide, and 7 meters high? _____

5. a) How many cubic yards of dirt must be removed for a basement 15 yards long, 8 yards wide, and 4 yards high? _____

b) How much will it cost to remove the dirt at $1.85 a cubic yard?

_____

6. The measurements for two boxes are shown below.

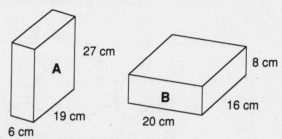

a) Which box contains more? _____

b) How much more? _____

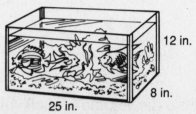

7. What is the volume of the aquarium?

_____

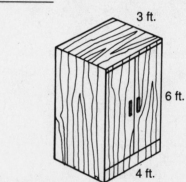

8. What is the volume of the storage cabinet?

_____

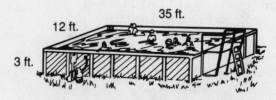

9. The swimming pool is filled to the top. How many cubic feet of water does it hold? _____

Choose the Formula

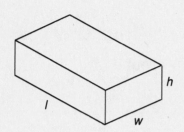

Rectangular Solid

$V = lwh$

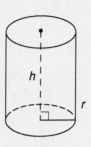

Cylinder

$V = \pi r^2 h$

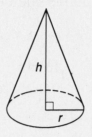

Cone

$V = \frac{1}{3} \pi r^2 h$

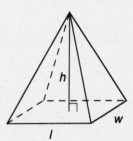

Rectangular Pyramid

$V = \frac{1}{3} lwh$

▶ Choose the correct formula; then find the volume. (Use 3.14 for π.)

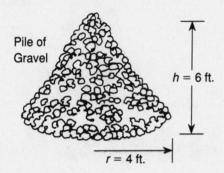

Pile of Gravel

$h = 6$ ft.

$r = 4$ ft.

1. $V =$ _____

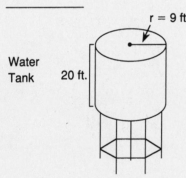

Water Tank   20 ft.

$r = 9$ ft.

2. $V =$ _____

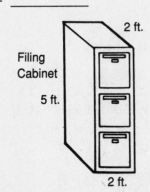

Filing Cabinet

2 ft.

5 ft.

2 ft.

3. $V =$ _____

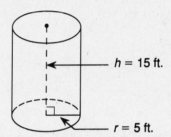

Grain Silo

$h = 15$ ft.

$r = 5$ ft.

4. $V =$ _____

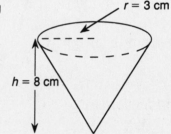

Drinking Cup

$r = 3$ cm

$h = 8$ cm

5. $V =$ _____

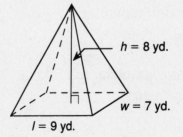

Pyramid

$h = 8$ yd.

$w = 7$ yd.

$l = 9$ yd.

6. $V =$ _____

Using Perimeter, Area, and Volume

Below is a brief review of perimeter, area, and volume.

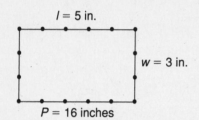

Perimeter is the distance completely around the outside of a figure. It is a measure of length.

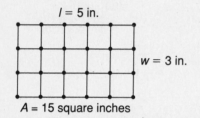

Area is the number of square units needed to exactly cover a surface. It is measured in square units.

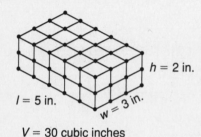

Volume is the number of cubic units needed to fill a space. It is measured in cubic units.

▶ Do you need to find the perimeter, area, or volume to do each of these tasks?

1. Filling a sandbox __volume__

2. Painting a room _____

3. Planting grass seed _____

4. Framing a picture _____

5. Buying material to make a pillowcase_____

6. Fencing a yard _____

7. Filling a swimming pool _____

8. Carpeting a floor _____

9. Adding braid to a tablecloth _____

10. Figuring out how much a storage bin holds _____

Review Your Skills

▶ Answer the following questions.

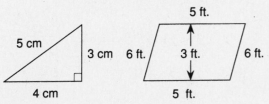

Perimeter = 16

Area = 14

1. Connect the dots to draw a figure for the perimeter and area above.

5 ft.

5 cm   3 cm   6 ft.   3 ft.   6 ft.

4 cm   5 ft.

a) $P =$ _____   **c)** $P =$ _____

b) $A =$ _____   **d)** $A =$ _____

2. Find the perimeter and area of each figure.

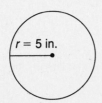

$r = 5$ in.

3. a) Find the area of the circle. _____

b) Find the circumference. _____

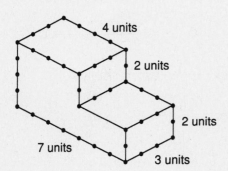

4 units

2 units

2 units

7 units

3 units

4. How many cubic units are contained in the solid? _____

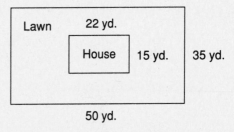

Lawn   22 yd.

House   15 yd.   35 yd.

50 yd.

5. a) What is the area of the lawn? _____

b) What is the perimeter of the house? _____

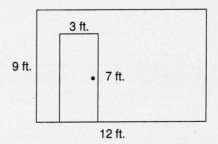

3 ft.

9 ft.   7 ft.

12 ft.

6. How many square feet of paneling will be needed to cover the wall, not including the door? _____

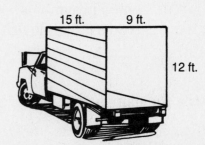

15 ft.   9 ft.

12 ft.

7. How many boxes 3 feet long, 3 feet wide, and 3 feet high will fit in the truck? Hint: Divide the volume of the truck by the volume of a box. _____

Same Shape — Same Size

Two figures are congruent if one figure can fit exactly on the other. **Congruent** means exactly alike in size and shape. Congruent figures do not have to be in the same position.

▶ Circle the letter of the figure that is congruent to the first figure.

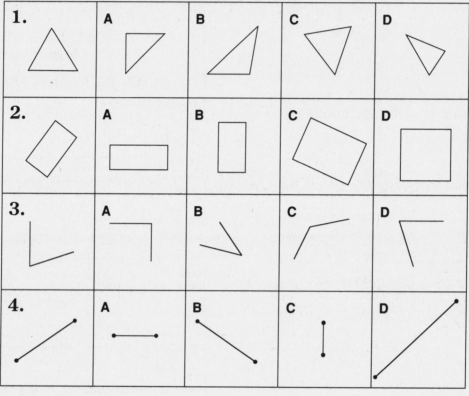

Triangles can be labeled using the triangle symbol (△) and the three vertex letters.

Example

The triangle at the right is represented in symbols as "△ ABC" and is read as triangle ABC.

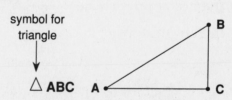

symbol for triangle

△ ABC

5. Write the names of the congruent triangles that are in the figure below.

△ _____ and △ _____

△ _____ and △ _____

△ _____ and △ _____

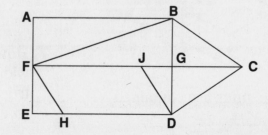

Drawing Congruent Figures

▶ Connect the dots to draw a congruent figure to the right of the one already drawn. The first one has been done for you.

1.
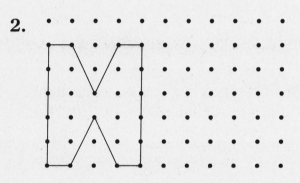

2.
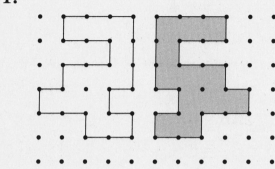

3.
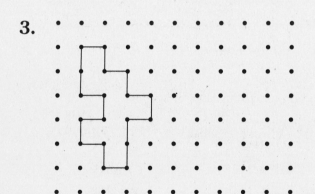

4.
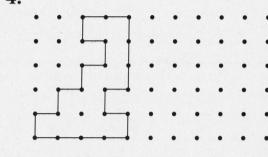

5.
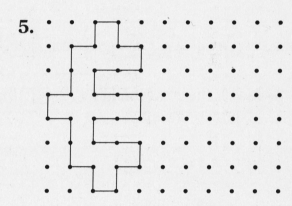

6.
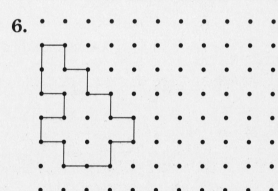

▶ Congruent figures do not have to be in the same position. Each figure below is congruent to a figure above. Write the number of the figure above that matches.

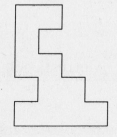

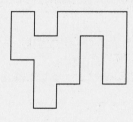

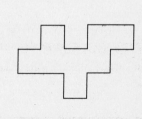

7. _____

8. _____

9. _____

10. _____

Corresponding Parts

Matching parts in congruent figures are called **corresponding parts**. When two figures are congruent, their corresponding parts are also congruent.

Example

symbol for congruent
↓
△ ABC ≅ △DEF

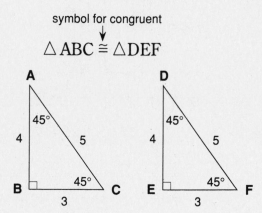

Corresponding Sides	Corresponding Angles
$\overline{AB} \cong \overline{DE}$	$\angle A \cong \angle D$
$\overline{BC} \cong \overline{EF}$	$\angle B \cong \angle E$
$\overline{AC} \cong \overline{DF}$	$\angle C \cong \angle F$

▶ Match the corresponding parts in the congruent triangles.

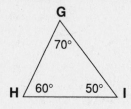

Corresponding Sides	Corresponding Angles

1. $\overline{GH} \cong \overline{JK}$ $\angle H \cong \angle K$

_____ ≅ _____   _____ ≅ _____

_____ ≅ _____   _____ ≅ _____

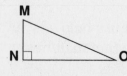

Corresponding Sides	Corresponding Angles

2. _____ ≅ _____   _____ ≅ _____

_____ ≅ _____   _____ ≅ _____

_____ ≅ _____   _____ ≅ _____

▶ Corresponding parts in congruent figures are the same size. Find the size of the corresponding parts.

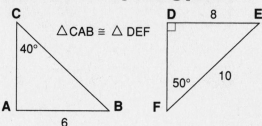

△CAB ≅ △ DEF

Corresponding Sides	Corresponding Angles

3. a) $\overline{AC}$ = ___8___   d) $\angle A$ = ___90°___

b) $\overline{DF}$ = _____   e) $\angle E$ = _____

c) $\overline{CB}$ = _____   f) $\angle B$ = _____

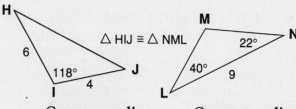

△ HIJ ≅ △ NML

Corresponding Sides	Corresponding Angles

4. a) $\overline{LM}$ = _____   d) $\angle M$ = _____

b) $\overline{HJ}$ = _____   e) $\angle J$ = _____

c) $\overline{MN}$ = _____   f) $\angle H$ = _____

Lines of Symmetry

A line on which you can fold a figure so the parts match up exactly is called a **line of symmetry**. The line will divide a figure into two congruent parts.

Examples

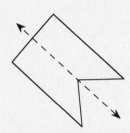  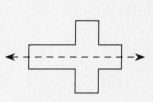

▶ Is the dotted line a line of symmetry?

  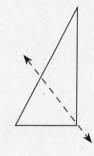 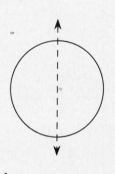

1. _____
   yes or no

2. _____
   yes or no

3. _____
   yes or no

4. _____
   yes or no

▶ Draw a line of symmetry for each figure.

5.

7.

9.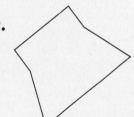

11.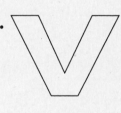

6.

8.

10.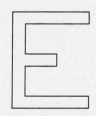

12.

Similar Figures

Figures that have the same shape but may or may not have the same size are called **similar figures.**

Example

These figures have the same shape but are a different size. They are similar figures.

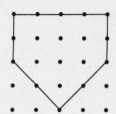

▶ Draw a similar figure that is twice as large as the one shown.

1.

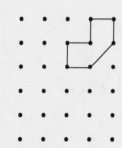

2.

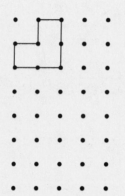

3.

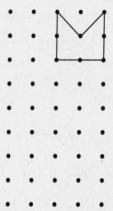

▶ Draw a similar figure that is one-half as large as the one shown.

4.

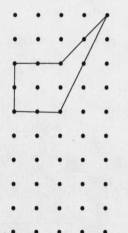

5.

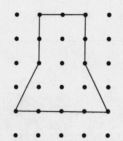

6.

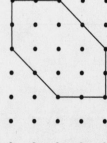

Corresponding Parts in Similar Figures

Matching parts in similar figures are also called corresponding parts. Enlarging or reducing a figure changes the lengths of the sides, but the sizes of the angles remain the same.

Example

symbol for similar

△ ABC ~ △DEF

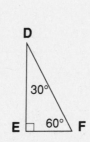

Corresponding Angles	Corresponding Sides
∠A = ∠D	$\overline{AB} \sim \overline{DE}$
∠B = ∠E	$\overline{BC} \sim \overline{EF}$
∠C = ∠F	$\overline{AC} \sim \overline{DF}$

▶ Write the corresponding angles and sides for the similar figures.

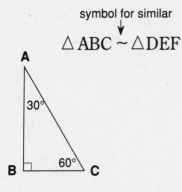

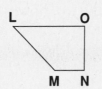

Corresponding Angles	Corresponding Sides
1. a) ∠A = ∠ _____	**d)** $\overline{AB}$ and _____
b) ∠B = ∠ _____	**e)** $\overline{AC}$ and _____
c) ∠C = ∠ _____	**f)** $\overline{BC}$ and _____

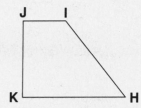

Corresponding Angles	Corresponding Sides
3. a) ∠O = ∠ _____	**e)** $\overline{NO}$ and _____
b) ∠M = ∠ _____	**f)** $\overline{LM}$ and _____
c) ∠L = ∠ _____	**g)** $\overline{OL}$ and _____
d) ∠N = ∠ _____	**h)** $\overline{MN}$ and _____

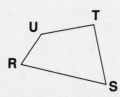

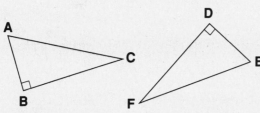

Corresponding Angles	Corresponding Sides
2. a) ∠R = ∠ _____	**e)** $\overline{RU}$ and _____
b) ∠U = ∠ _____	**f)** $\overline{ST}$ and _____
c) ∠T = ∠ _____	**g)** $\overline{RS}$ and _____
d) ∠S = ∠ _____	**h)** $\overline{UT}$ and _____

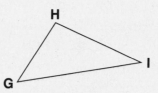

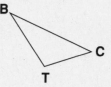

Corresponding Angles	Corresponding Sides
4. a) ∠G = ∠ _____	**d)** $\overline{GH}$ and _____
b) ∠I = ∠ _____	**e)** $\overline{GI}$ and _____
c) ∠H = ∠ _____	**f)** $\overline{HI}$ and _____

Using Proportions

The lengths of the corresponding sides of similar figures are proportional. In the example below, the sides of $\triangle AGH$ are all twice the size of the sides of $\triangle BCD$. You can use proportions to find the unknown lengths of similar figures.

Example: Find the length of $\overline{CD}$.

Set up and solve a proportion using the lengths of the corresponding sides. Let $\overline{CD} = n$.

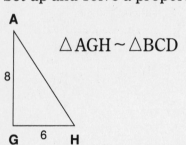 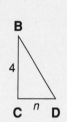

$$\triangle AGH \sim \triangle BCD$$

larger figure $\dfrac{8}{4} = \dfrac{6}{n}$ larger figure
smaller figure $\phantom{\dfrac{8}{4}}$ smaller figure

$8 \times n = 6 \times 4$ Cross products are equal.

$\dfrac{8 \times n}{8} = \dfrac{24}{8}$ Divide both sides by 8.

$n = 3$ The length of $\overline{CD}$ is 3.

▶ Use a proportion to find the unknown length (n) for each similar figure.

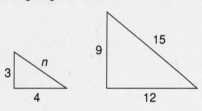

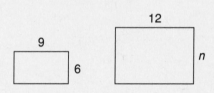

1. $\dfrac{\square}{\square} = \dfrac{\square}{\square}$

      $n = $ _____

3. $\dfrac{\square}{\square} = \dfrac{\square}{\square}$

      $n = $ _____

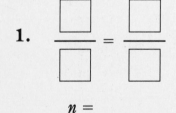

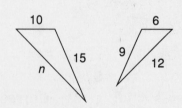

2. $\dfrac{\square}{\square} = \dfrac{\square}{\square}$

      $n = $ _____

4. $\dfrac{\square}{\square} = \dfrac{\square}{\square}$

      $n = $ _____

Similar Triangles to Measure

Many problems can be solved using similar triangles. You can use a proportion to help you find the length of a missing side in one of two similar triangles.

Example: Find the length of $\overline{DE}$.

Set up and solve a proportion using the lengths of the corresponding sides. Let $\overline{DE} = n$.

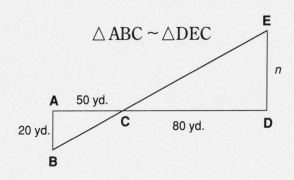

$\triangle ABC \sim \triangle DEC$

smaller triangle $\dfrac{50}{80} = \dfrac{20}{n}$ smaller triangle
larger triangle   larger triangle

$$50 \times n = 20 \times 80$$

$$\frac{5\!\!\!/0 \times n}{5\!\!\!/0} = \frac{1{,}600}{50}$$

$$n = 32$$

The length of $\overline{DE}$ is 32 yards.

▶ Set up a proportion to find the unknown length. Let n represent the unknown length.

1. Find the length of $\overline{XW}$.
   $\triangle TUV \sim \triangle TXW$

$$\frac{\Box}{\Box} = \frac{\Box}{\Box}$$

$n =$ _____

3. Find the length of $\overline{XP}$.
   $\triangle TRS \sim \triangle XPS$

$$\frac{\Box}{\Box} = \frac{\Box}{\Box}$$

$n =$ _____

2. Find the length of $\overline{GH}$.
   $\triangle DEF \sim \triangle GHI$

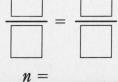

$$\frac{\Box}{\Box} = \frac{\Box}{\Box}$$

$n =$ _____

4. Find the length of $\overline{GD}$.
   $\triangle FCE \sim \triangle FGD$

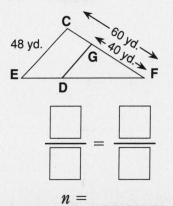

$$\frac{\Box}{\Box} = \frac{\Box}{\Box}$$

$n =$ _____

Similar Triangles and Indirect Measurement

Using proportions can be a great help in finding distances you cannot measure directly.

▶ Use a proportion to solve the problems below.

1. Using a pole and shadows, find the height of the tree.

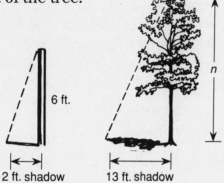

6 ft.

2 ft. shadow   13 ft. shadow

$$\frac{\boxed{}}{\boxed{}} = \frac{\boxed{}}{\boxed{}}$$

$n = $ _____

3. Find the distance across the pond.

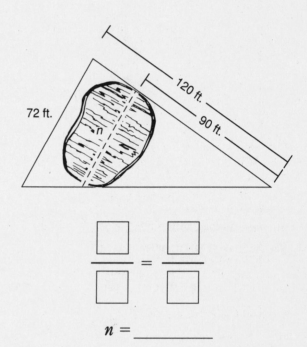

72 ft.

120 ft.

90 ft.

$$\frac{\boxed{}}{\boxed{}} = \frac{\boxed{}}{\boxed{}}$$

$n = $ _____

2. What is the distance across the lake?

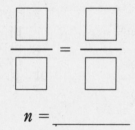

12 yd.

40 yd.

9 yd.

$$\frac{\boxed{}}{\boxed{}} = \frac{\boxed{}}{\boxed{}}$$

$n = $ _____

4. Use the short tree to find the height of the tall tree.

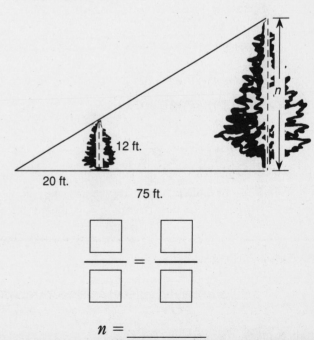

12 ft.

20 ft.

75 ft.

$$\frac{\boxed{}}{\boxed{}} = \frac{\boxed{}}{\boxed{}}$$

$n = $ _____

ANSWER KEY

Page 1: Geometric Shapes

Compare your answers with other students' answers.

Page 2: Geometric Ideas

1. Points: G, B, A, C, D, E, F
2. Line segments: $\overline{GE}$, $\overline{EF}$, $\overline{AB}$, $\overline{CD}$
3. Lines: $\overleftrightarrow{AB}$, $\overleftrightarrow{CD}$
4. Points: K, L, M, P, B, C
5. Line segments: $\overline{KL}$, $\overline{LM}$, $\overline{BC}$, $\overline{KM}$
6. Points: A, B, C, D, E, F
7. $\overline{AB}$, $\overline{CD}$, $\overline{DE}$, $\overline{EF}$, $\overline{FC}$

Page 3: Angles

1. ∠CDE or ∠EDC
2. ∠GFE or ∠EFG
3. ∠PCE or ∠ECP
4. a) ∠ANP   d) ∠PNX
   b) ∠ANC   e) ∠CNX
   c) ∠CNP   f) ∠ANX
5. Answers may vary. Possible answers are:

   ∠CAD, ∠CAE, ∠AEB, ∠BEH, ∠AEH,
   ∠EBK, ∠KFG, ∠EHG, ∠HGF, ∠FKB

Page 4: Geometric Models

1. a) A, B, C, D, E, F, G
   b) $\overline{CD}$, $\overline{AB}$, $\overline{EG}$, $\overline{EF}$
   c) ∠FEG
   d) $\overleftrightarrow{AB}$, $\overleftrightarrow{CD}$
2. Point
3. a) X, Y, Z, E
   b) $\overline{XY}$, $\overline{YZ}$
   c) ∠XYZ
4. a) K, P, B, L, C, X, M, Y
   b) $\overline{KL}$, $\overline{LM}$, $\overline{BC}$, $\overline{XY}$, $\overline{KM}$
   c) ∠KLM, ∠LMK, ∠MKL
5. a) N, O, P, D
   b) $\overline{ON}$, $\overline{OP}$, $\overline{OD}$
   c) ∠NOP, ∠POD, ∠NOD
6. a) A, C, E, F, B, J, P, D
   b) $\overline{AC}$, $\overline{CE}$, $\overline{FJ}$, $\overline{BD}$
   c) ∠ACE
   d) $\overrightarrow{BD}$

Page 5: Size of Angles

1. b   5. a   9. ⌐
2. a   6. a
3. b   7. a and b   10. ∠
4. b   8. a and c   11. ∠

Page 6: Using a Protractor

1. a) Inside
   b) 120°
2. a) Outside
   b) 140°

Page 7: Measuring Angles

Answers may vary by a few degrees:

1. 68°   5. 145°   8. 90°
2. 35°   6. 25°   9. 45°
3. 110°   7. 155°   10. 130°
4. 60°

Page 8: Classifying Angles

1. a) =
   b) Right angle
2. a) >
   b) Obtuse angle
3. a) =
   b) Right angle
4. a) <
   b) Acute angle
5. a) About 45°
   b) Acute angle
6. a) About 90°
   b) Right angle
7. a) About 120°
   b) Obtuse angle
8. a) About 160°
   b) Obtuse angle
9. Acute angle
10. Obtuse angle
11. Right angle
12. Acute angle
13. Obtuse angle
14. Acute angle

Page 9: Complementary Angles

1. 30°   4. 5°   7. 61°
2. 45°   5. 44°   8. 30°
3. 65°   6. 82°   9. 37°

10. a)   b)   c)   d)

Page 10: Supplementary Angles

1. 130°   4. 160°   7. 11°
2. 40°   5. 26°   8. 135°
3. 90°   6. 95°   9. 75°

10. a) b) c)   d)

Page 11: Vertical, Horizontal, and Slanting Lines

1. Answers may vary:

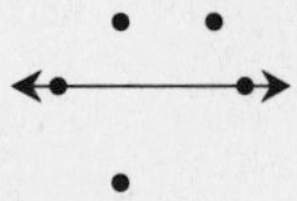

2. Answers may vary:

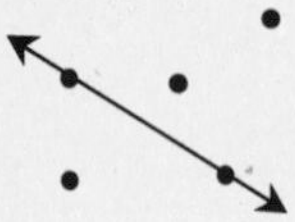

3.

4. a) $\overrightarrow{AN}$

b) $\overleftrightarrow{LB}$

c) $\overrightarrow{PK}$

5. a) $\overleftrightarrow{BH}$

b) $\overleftrightarrow{AC}$

c) $\overleftrightarrow{EJ}$, $\overrightarrow{KF}$

Page 12: Parallel and Intersecting Lines

1. **a)** C   **b)** Y   **c)** T   **d)** I
2. **a)** Intersecting
   b) Parallel
   c) Intersecting
   d) Parallel
3. **a)** 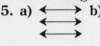   **b)**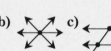
4. Intersecting
5. **a)** **b)** **c)** 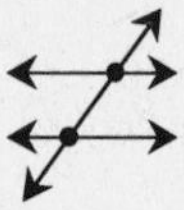 **d)**

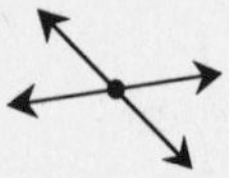

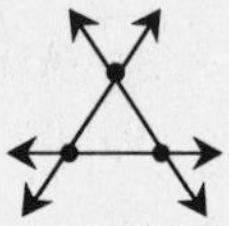

Page 13: Perpendicular Lines

1. **a)** Perpendicular
   b) Intersecting
   c) Intersecting
   d) Perpendicular

2.   3. 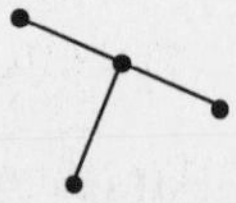   4.

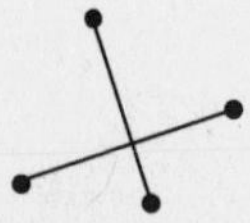

5. Answers may vary. Eight possible answers are:

a) $\overline{GF}$ and $\overline{BD}$ **e)** $\overline{FB}$ and $\overline{BA}$
b) $\overline{FG}$ and $\overline{GH}$ **f)** $\overline{BA}$ and $\overline{AC}$
c) $\overline{GH}$ and $\overline{HE}$ **g)** $\overline{AC}$ and $\overline{CD}$
d) $\overline{FE}$ and $\overline{HE}$ **h)** $\overline{CD}$ and $\overline{DB}$

Page 14: Street Plans

1. **a)** Kennedy and Wilson
   b) Adam and Wilson
   c) Kennedy and Lincoln

Page 14: Street Plans (continued)

2. Wilson and Lincoln
3. **a)** Kennedy and Wilson
   b) Kennedy and Lincoln
4. Thompson and Jackson
5. No
6. **a)** Thompson and Miller
   b) Thompson and Drew
7. Walnut
8. Park
9. Angling and Park
10. **a)** Oak and Beech
    b) Beech and Elm
11. **a)** Pine and Maple
    b) Elm and Oak
12. Beech, Pine, and Maple

Page 15: Review Your Skills

1. **a)** 150°
   b) 50°
   c) 90°
   d) 130°
2. **a)** 90°
   b) Right angle
3. **a)** 15°
   b) Acute angle
4. **a)** 165°
   b) Obtuse angle
5. **a)** 45°
   b) Acute angle

6. **a)** 135°
   b) Obtuse angle
7. **a)** $\overleftrightarrow{YX}$
   b) $\overrightarrow{EG}$
   c) $\angle HFJ$
8. Complementary angles
9. Supplementary angles
10. Right angle
11. **a)** $\overrightarrow{AB}$ and $\overleftrightarrow{CD}$
    $\overleftrightarrow{AB}$ and $\overrightarrow{EF}$
    b) $\overline{CD}$ and $\overline{EF}$

Page 16: Drawing Polygons

1. 8
2. 7
3. 5
4. 12
5. 3
6. 6
7. 4
8. 5

Drawings may vary:

9.   10. 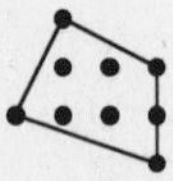   11. 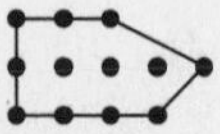   12. 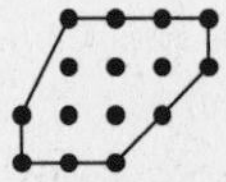   13. 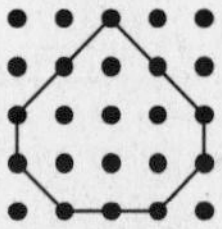   14.

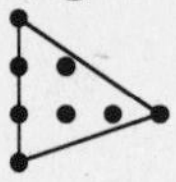

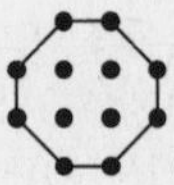

Page 17: Working with Polygons

1. B
2. A
3. C

Arrangements may vary:

4.   5.

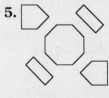

Page 18: Regular Polygons

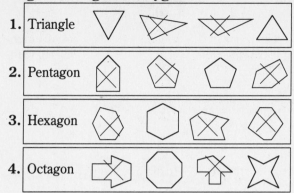

1. Triangle
2. Pentagon
3. Hexagon
4. Octagon

5. Not regular
6. Regular
7. Regular
8. Not regular

Page 19: Special Quadrilaterals

1. Trapezoid
2. Square
3. Parallelogram
4. Rectangle

5. Rectangle
6. Parallelogram
7. Trapezoid
8. Square

Page 20: Parallelograms

Drawings may vary:

1.
2.
3.
4.
5.
6.

Page 21: Triangles Classified by Sides

1. a) Equilateral triangle
   b) Scalene triangle
   c) Isosceles triangle
2. a) Scalene triangle
   b) Isosceles triangle
   c) Equilateral triangle

Drawings may vary:

3.
4.
5.

Page 22: Triangles Classified by Angles

1. a) Right triangle
   b) Obtuse triangle
   c) Acute triangle
   d) Obtuse triangle
2. a) Acute triangle
   b) Right triangle
   c) Obtuse triangle

Drawings may vary:

3.
4.
5.

Page 23: The Sum of Angles in a Triangle

1. $\angle J = 25°$
2. $\angle E = 48°$
3. $\angle N = 51°$
4. $\angle T = 50°$
5. $\angle D = 19°$
6. $\angle B = 38°$

Page 24: Review of Polygons and Triangles

1. B, D, G
2. B
3. B, G
4. A, E
5. B, C, D, G
6. F, I
7. A, B, F, H
8. C
9. H
10. C
11. B
12. A
13. D
14. a) $\angle A = 40°$
    b) $\angle H = 130°$
    c) $\angle D = 35°$

Page 25: Tiling Floors

1. B
2. A
3. D
4. C
5. C
6. D
7. A
8. B

Drawings may vary:

9.
10.
11.
12.

Page 26: Measure of Length

1. 5 units
2. 9 units
3. 17 units
4. 17 units
5. 12 units

Page 27: Distance Around

1. 18 units
2. 26 units
3. 30 units

Drawings may vary:

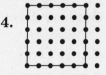

4.

Page 27: Distance Around (continued)

5. 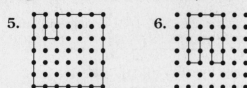 6.

Page 28: Find the Perimeter

1. 154 units   3. 171 units   5. 153 units
2. 142 units   4. 146 units   6. 187 units

Page 29: Missing Lengths of Polygons

1. 5 feet   4. 18 meters   7. 72 centimeters
2. 2 meters   5. 168 inches   8. 288 yards
3. 35 yards   6. 100 feet

Page 30: Perimeter Formulas

1. a) $P = 4s$
   b) $P = 4 \times 7$
   c) $P = 28$ inches

2. a) $P = a + b + c$
   b) $P = 19 + 17 + 21$
   c) $P = 57$ centimeters

3. a) $P = a + b + c$
   b) $P = 8 + 9 + 7$
   c) $P = 24$ feet

4. a) $P = 2l + 2w$
   b) $P = (2 \times 12) + (2 \times 5)$
   c) $P = 34$ meters

5. a) $P = 2l + 2w$
   b) $P = (2 \times 15) + (2 \times 7)$
   c) $P = 44$ centimeters

6. a) $P = 4s$
   b) $P = 4 \times 19$
   c) $P = 76$ yards

Page 31: Perimeter Applications

1. 138 feet   4. 48 inches   6. 96 inches
2. 80 yards   5. 560 yards   7. 120 feet
3. 240 feet

Page 32: Applying Perimeter Skills

1. a) $P = 2l + 2w$
   b) $P = (2 \times 28) + (2 \times 18)$
   c) $P = 92$ feet

Page 32: Applying Perimeter Skills (continued)

2. a) $P = 4s$ 4. 21 inches
   b) $P = 4 \times 85$ 5. 10 yards
   c) $P = 340$ yards   6. 17 feet

3. a) $P = 2l + 2w$
   b) $P = (2 \times 25) + (2 \times 12)$
   c) $P = 74$ inches

Page 33: Surface Inside a Figure

1. a) 2 units   3. a) 3 units
   b) 4 units   b) 5 units
   c) 3 units   c) 4 units

2. a) 3 units   4. a) 5 units
   b) 4 units   b) 6 units
   c) 8 units   c) 7 units

Page 34: Square Units

1. 9 square units   6. 15 square units
2. 15 square units   7. 15 square units
3. 21 square units   8. 17 square units
4. 16 square units   9. 23 square units
5. 13 square units

Page 35: Square Inches and Square Centimeters

1. 5 square inches   4. 14 square centimeters
2. 6 square inches   5. 19 square centimeters
3. 13 square centimeters

Page 36: Rectangles

1. 12 square feet
2. 648 square inches
3. 289 square centimeters
4. 2,560 square yards
5. 2,304 square meters
6. 2,080 square centimeters

Page 37: Parallelograms

1. 1,247 square inches
2. 242 square centimeters
3. 1,015 square yards
4. 765 square meters
5. 1,292 square inches
6. 3,450 square feet

Page 38: Triangles

1. a) $A = \frac{1}{2}bh$
   b) $A = \frac{1}{2} \times 18 \times 20$
   c) $A = 180$ square centimeters

2. a) $A = \frac{1}{2}bh$
   b) $A = \frac{1}{2} \times 55 \times 22$
   c) $A = 605$ square yards

3. a) $A = \frac{1}{2}bh$
   b) $A = \frac{1}{2} \times 56 \times 13$
   c) $A = 364$ square meters

4. a) $A = \frac{1}{2}bh$
   b) $A = \frac{1}{2} \times 33 \times 28$
   c) $A = 462$ square inches

Page 39: Practice Your Skills

1. a) $A = \frac{1}{2}bh$
   b) 27 square inches

2. a) $A = lw$
   b) 625 square yards

3. a) $A = bh$
   b) 216 square meters

4. a) $A = \frac{1}{2}bh$
   b) 260 square inches

5. a) $A = \frac{1}{2}bh$
   b) 972 square centimeters

6. a) $A = bh$
   b) 1,116 square inches

7. a) $A = lw$
   b) 912 square centimeters

8. a) $A = lw$
   b) 5,625 square yards

Page 40: Dividing Figures into Parts

Dotted lines may vary in 1–4:

1.

2.

3.

4.

5. 104 square centimeters
6. 264 square inches
7. 98 square meters
8. 1,813 square feet
9. 112 square yards
10. 5,684 square centimeters

Page 41: Drawing to Find Areas and Perimeters

Drawings may vary:

1. 10 square units

6. 18 units

2. $21\frac{1}{2}$ square units

7.
Perimeter = 14 Area = 10

3. $8\frac{1}{2}$ square units

8.
Perimeter = 18 Area = 20

4. 14 units

9.
Perimeter = 16 Area = 15

5. 22 units

Page 42: Perimeter and Area Review

1. C, B, A
2. B, D, C, A
3. 6 tiles
4. a) 136 centimeters
   b) 96 feet
   c) 88 inches
   d) 106 meters
5. a) 36 square inches
   b) 1,485 square yards
   c) 336 square feet
   d) 570 square inches

6. a) $P = 12$ meters, $A = 6$ square meters
   b) $P = 102$ feet, $A = 561$ square feet
   c) $P = 80$ meters, $A = 225$ square meters
   d) $P = 36$ meters, $A = 81$ square meters

Page 43: Using Perimeter and Area

1. a) Perimeter
   b) 42 yards
2. a) Area
   b) 80 square yards

3. a) Area
   b) 54 square feet
4. a) Perimeter
   b) 36 feet

Page 43: Using Perimeter and Area (continued)

5. a) Area
  b) 162 tiles
6. a) Perimeter
  b) 54 feet

7. a) Area
  b) 15 square yards
8. a) Perimeter
  b) 17 yards

Page 44: Perimeter and Area Applications

1. a) 3 pounds
  b) $7.35
2. a) Yes
  b) 300 feet
3. a) $154.05
  b) 32 feet

4. 320 feet
5. No
6. 396 centimeters
7. $382.80

Page 45: Reading House Plans

1. a) 63 feet
  b) 39 feet
  c) 39 feet
  d) 63 feet
  e) 51 feet
  f) 51 feet

2. $168.30
3. $815.85
4. $880
5. a) $602
  b) $430
  c) $430

Page 46: Learning about Circles

1. a) 56 feet
  b) 13 inches
  c) 30 centimeters
  d) 6 meters

2. a) 8 inches
  b) 15 centimeters
  c) 4 yards
  d) 22 meters

Page 47: Finding the Circumference

1. a) $C = \pi d$
  b) $C = 3.14 \times 5$
  c) $C = 15.7$ feet

2. a) $C = \pi d$
  b) $C = 3.14 \times 25$
  c) $C = 78.5$ inches

3. a) $C = \pi d$
  b) $C = 3.14 \times 9.5$
  c) $C = 29.83$ meters

4. a) $C = \pi d$
  b) $C = 3.14 \times 14$
  c) $C = 43.96$ centimeters

5. a) $C = \pi d$
  b) $C = 3.14 \times 12$
  c) $C = 37.68$ centimeters

6. a) $C = \pi d$
  b) $C = 3.14 \times 6$
  c) $C = 18.84$ yards

Page 47: Finding the Circumference (continued)

7. a) $C = \pi d$
  b) $C = 3.14 \times 50$
  c) $C = 157$ centimeters

8. a) $C = \pi d$
  b) $C = 3.14 \times 9$
  c) $C = 28.26$ inches

Page 48: Area of a Circle

1. a) $A = \pi r^2$
  b) $A = 3.14 \times 5 \times 5$
  c) $A = 78.5$ square centimeters

2. a) $A = \pi r^2$
  b) $A = 3.14 \times 12 \times 12$
  c) $A = 452.16$ square inches

3. a) $A = \pi r^2$
  b) $A = 3.14 \times 65 \times 65$
  c) $A = 13{,}266.5$ square meters

4. a) $A = \pi r^2$
  b) $A = 3.14 \times 4 \times 4$
  c) $A = 50.24$ square millimeters

5. a) $A = \pi r^2$
  b) $A = 3.14 \times 11 \times 11$
  c) $A = 379.94$ square centimeters

6. a) $A = \pi r^2$
  b) $A = 3.14 \times 20 \times 20$
  c) $A = 1{,}256$ square inches

Page 49: Using Circles

1. $3.14 \times (20)^2 = 1{,}256$
1,256 square feet

2. $3.14 \times 2(8) = 50.24 =$
50 feet

3. $3.14 \times (5)^2 = 78.5 =$
79 square meters

4. $3.14 \times 25 = 78.5 =$
79 feet

5. $3.14 \times (\frac{15}{2})^2 = 176.625$
177 square meters

6. $3.14 \times 6 = 18.84$
18.84 feet

7. $3.14 \times (6)^2 = 113.04$
$3.14 \times (5)^2 = 78.5$
$113.04 - 78.5 = 34.54$
34.54 square meters

8. $64 - [3.14 \times (4)^2] =$
$64 - 50.24 = 13.76$
13.76 square centimeters

9. $3.14 \times (20)^2 = 1{,}256$
$2(3.14 \times 10^2) = 628$
$1{,}256 - 628 = 628 =$
628 square millimeters

10. $18^2 - [3.14 \times (9)^2] =$
$324 - 254.34 = 69.66$
69.66 square centimeters

Page 50: Count the Cubes

1. 6 cubes
2. 12 cubes
3. 11 cubes
4. 9 cubes
5. 12 cubes

Page 51: Relating a Solid to a Drawing

1. 6 cubic units
2. 16 cubic units
3. 9 cubic units
4. 7 cubic units
5. 14 cubic units
6. 10 cubic units
7. 11 cubic units
8. 17 cubic units

Page 52: Hidden Cubes

1. 12 cubic units
2. 30 cubic units
3. 9 cubic units
4. 18 cubic units
5. 19 cubic units
6. 30 cubic units

Page 53: Finding Volume

1. a) $V = 5 \times 2 \times 2$
   b) $V = 20$ cubic units
2. a) $V = 4 \times 4 \times 4$
   b) $V = 64$ cubic units
3. a) $V = 4 \times 3 \times 3$
   b) $V = 36$ cubic units
4. a) $V = 4 \times 3 \times 7$
   b) $V = 84$ cubic units

Page 54: Apply Your Skills

1. 63 cubic centimeters
2. 21 cubic feet
3. 1,620 cubic inches
4. 125 cubic yards
5. 168 cubic yards
6. 540 cubic inches
7. 200 cubic yards
8. 936 cubic centimeters

Page 55: Two-Step Volume Problems

1. 112 cubic units
2. 152 cubic units
3. 90 cubic units
4. 300 cubic units

Page 56: Volume Applications

1. 750 cubic feet
2. 360 cubic meters
3. 36 cubic feet
4. 1,400 cubic meters
5. a) 480 cubic yards
   b) $888
6. a) Box A
   b) 518 cubic centimeters
7. 2,400 cubic inches
8. 72 cubic feet
9. 1,260 cubic feet

Page 57: Choose the Formula

1. $\frac{1}{3} (3.14) (4)^2 (6) = 100.48$ cubic feet
2. $3.14 (9)^2 (20) = 5,086.8$ cubic feet
3. $2 \times 2 \times 5 = 20$ cubic feet
4. $3.14 (5)^2 (15) = 1,177.5$ cubic feet
5. $\frac{1}{3} (3.14) (3)^2 (8) = 75.36$ cubic centimeters
6. $\frac{1}{3} (9) (7) (8) = 168$ cubic yards

Page 58: Using Perimeter, Area, and Volume

1. Volume
2. Area
3. Area
4. Perimeter
5. Area
6. Perimeter
7. Volume
8. Area
9. Perimeter
10. Volume

Page 59: Review Your Skills

1. Drawing may vary:

2. a) $P = 12$ centimeters
   b) $A = 6$ square centimeters.
   c) $P = 22$ feet
   d) $A = 15$ square feet
3. a) 78.5 square inches
   b) 31.4 inches
4. 66 cubic units
5. a) 1,420 square yards
   b) 74 yards
6. 87 square feet
7. 60 boxes

Page 60: Same Shape—Same Size

1. c
2. b
3. d
4. b
5. △ABF and △GFB
   △BGC and △DGC
   △FEH and △DGJ

Page 61: Drawing Congruent Figures

1–6: Compare your answers with other students' answers.
7. 4
8. 6
9. 1
10. 3

Page 62: Corresponding Parts

Sides:	Angles:
1. $\overline{GH} \cong \overline{JK}$	$\angle H \cong \angle K$
$\overline{GI} \cong \overline{JL}$	$\angle G \cong \angle J$
$\overline{HI} \cong \overline{KL}$	$\angle I \cong \angle L$

Sides:	Angles:
2. $\overline{MN} \cong \overline{PQ}$	$\angle N \cong \angle Q$
$\overline{NO} \cong \overline{QR}$	$\angle M \cong \angle P$
$\overline{MO} \cong \overline{PR}$	$\angle O \cong \angle R$

3. a) $\overline{AC} = 8$
   b) $\overline{DF} = 6$
   c) $\overline{CB} = 10$
   d) $\angle A = 90°$
   e) $\angle E = 40°$
   f) $\angle B = 50°$

4. a) $\overline{LM} = 4$
   b) $\overline{HJ} = 9$
   c) $\overline{MN} = 6$
   d) $\angle M = 118°$
   e) $\angle J = 40°$
   f) $\angle H = 22°$

Page 63: Lines of Symmetry

1. Yes
2. No
3. No
4. Yes

5.

6.

7.

8.

9.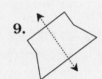

10.

11.

12.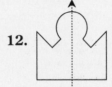

Page 64: Similar Figures

1.
2.
3.
4.
5.
6.

Page 65: Corresponding Parts in Similar Figures

1. a) $\angle A = \angle E$
    b) $\angle B = \angle D$
    c) $\angle C = \angle F$
    d) $\overline{AB}$ and $\overline{DE}$
    e) $\overline{AC}$ and $\overline{EF}$
    f) $\overline{BC}$ and $\overline{DF}$

2. a) $\angle R = \angle E$
    b) $\angle U = \angle F$
    c) $\angle T = \angle G$
    d) $\angle S = \angle H$
    e) $\overline{RU}$ and $\overline{EF}$
    f) $\overline{ST}$ and $\overline{HG}$
    g) $\overline{RS}$ and $\overline{EH}$
    h) $\overline{UT}$ and $\overline{FG}$

3. a) $\angle O = \angle K$
    b) $\angle M = \angle I$
    c) $\angle L = \angle H$
    d) $\angle N = \angle J$
    e) $\overline{NO}$ and $\overline{JK}$
    f) $\overline{LM}$ and $\overline{HI}$
    g) $\overline{OL}$ and $\overline{KH}$
    h) $\overline{MN}$ and $\overline{IJ}$

4. a) $\angle G = \angle C$
    b) $\angle I = \angle B$
    c) $\angle H = \angle T$
    d) $\overline{GH}$ and $\overline{CT}$
    e) $\overline{GI}$ and $\overline{CB}$
    f) $\overline{HI}$ and $\overline{TB}$

Page 66: Using Proportions

Proportion setups may vary; the solution for n will be the same:

1. $\frac{9}{3} = \frac{15}{n}$
    $n = 5$

2. $\frac{10}{15} = \frac{20}{n}$
    $n = 30$

3. $\frac{9}{12} = \frac{6}{n}$
    $n = 8$

4. $\frac{9}{15} = \frac{12}{n}$
    $n = 20$

Page 67: Similar Triangles to Measure

Proportion setups may vary:

1. $\frac{16}{40} = \frac{12}{n}$
    $n = 30$ yards

2. $\frac{3}{6} = \frac{4}{n}$
    $n = 8$ miles

3. $\frac{8}{24} = \frac{6}{n}$
    $n = 18$ feet

4. $\frac{60}{40} = \frac{48}{n}$
    $n = 32$ yards

Page 68: Similar Triangles and Indirect Measurement

Proportion setups may vary:

1. $\frac{2}{13} = \frac{6}{n}$
    $n = 39$ feet

2. $\frac{12}{40} = \frac{9}{n}$
    $n = 30$ yards

3. $\frac{120}{90} = \frac{72}{n}$
    $n = 54$ feet

4. $\frac{20}{75} = \frac{12}{n}$
    $n = 45$ feet